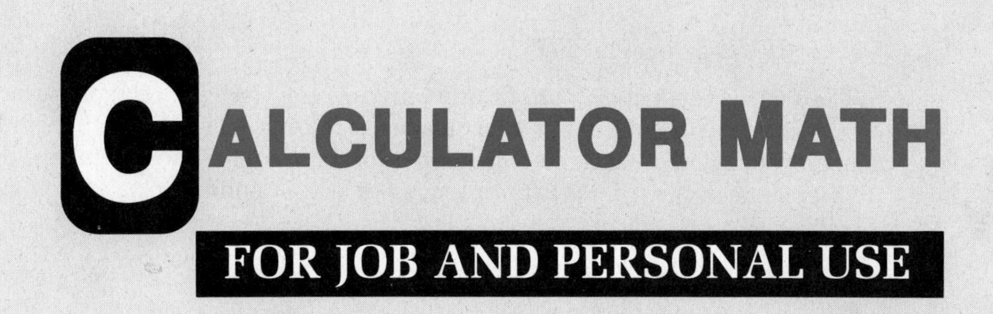

# CALCULATOR MATH

## FOR JOB AND PERSONAL USE

**William R. Pasewark**
Office Management Consultants
and
Professor Emeritus
Texas Tech University
Lubbock, Texas

**Merle Wood**
Education Consultant
Formerly of Oakland Public Schools
Lafayette, California

***South-Western Publishing Co.***

ISBN: 0-538-70481-0

1 2 3 4 5 6 7 8 9 0 DH 0 9 8 7 6 5 4 3 2 1

Printed in the United States of America

Acquisitions Editor: Karen Schneiter
Series Editor: Mark Linton
Developmental Editor: Gayle Entrup
Production Editor: Martha G. Conway
Designer: Darren Wright
Production Artist: Steve McMahon
Associate Photo Editor/Stylist: Linda Ellis
Marketing Manager: Shelly Battenfield

South-Western Publishing Co. gratefully acknowledges the foresight and commitment that Ben Willard, Acquisitions Editor, gave to this LIFE Series.

PHOTO CREDITS

Page 24, left: Monroe Systems for Business
Page 61: Photo courtesy of Remanco Systems, Inc.
Page 75: Courtesy of NCR Corporation
Page 94: Photo courtesy of SHARP ELECTRONICS CORPORATION

Library of Congress Cataloging-in-Publication Data

Pasewark, William Robert.
Calculator math for job and personal use / William R. Pasewark, Merle Wood.
p. cm.
Includes index.
ISBN 0-538-70481-0
1. Calculators. 2. Arithmetic—Data processing. I. Wood, Merle W. II. Title.
QA75.P36 1992
510′.28—dc20 91-26868
CIP

# PREFACE

Basic skills are required for each of us to conduct our personal and business dealings. An increasing need exists to provide adults with these basic skills so they can improve both their personal interactions and employment opportunities.

As a result, the LIFE Series was developed because South-Western believes that **L**earning **I**s **F**or **E**veryone (**LIFE**). The LIFE Series is specifically designed to provide adults with the basic skills needed for personal dealings and for job opportunities.

## THE LIFE SERIES

The LIFE Series is a self-paced, competency-based program specifically designed for adults to develop basic skills for job and personal use. Each book in the Series provides interesting material, realistic examples, practical applications, and flexible instruction to promote learner success and self-confidence.

The LIFE Series is divided into three basic skill areas: communication skills, math skills, and life skills. *Calculator Math for Job and Personal Use* is one of the math skills books in the LIFE Series. Each text–workbook is complete and may be used individually or in a series. Following is a complete list of the LIFE Series.

COMMUNICATION SKILLS

- *Spelling for Job and Personal Use*
- *Reading for Job and Personal Use*
- *Grammar and Writing for Job and Personal Use*
- *Punctuation, Capitalization, and Handwriting for Job and Personal Use*
- *Listening and Speaking for Job and Personal Use*

MATH SKILLS

- *Basic Math for Job and Personal Use*
- *Decimals, Fractions, and Percentages for Job and Personal Use*
- *Calculator Math for Job and Personal Use*

LIFE SKILLS

- *Career Planning and Development*
- *Problem Solving and Decision Making*
- *Self-Esteem and Getting Ahead*
- *Money Management*
- *Finding and Holding a Job*

## STRUCTURE AND ORGANIZATION

Each book in the LIFE Series has the same appearance and structure, enabling learners to experience more success and gain self-confidence as they progress. Competency-based instruction is also used throughout by presenting clear objectives followed by short segments of material, and then by specific exercises for immediate reinforcement.

The organization of *Calculator Math for Job and Personal Use* reflects the emphasis on learning to use the calculator in an everyday setting. In Part One, Basic Math with Your Calculator, students review the four basic functions of math. The focus of Part Two, Combined and Memory Operations, is on learning to combine the four basic math operations and to utilize the memory function on a hand-held calculator. In Part Three, Fractions, Decimals, and Percents, the student learns to use the basic math functions with fractions, decimals, and percents. Part Four, Business Forms Simulation, consists of a work situation in which a student employs all of the techniques learned in this book. Students are evaluated at each level.

The Glossary, Index, Answers, and Personal Progress Record at the end of *Calculator Math for Job and Personal Use* are designed to facilitate and enhance independent student learning and achievement.

## SPECIAL FEATURES OF *CALCULATOR MATH FOR JOB AND PERSONAL USE*

*Calculator Math for Job and Personal Use* is a complete and comprehensive package providing the student with learning material written specifically to meet the unique needs of the adult learner and providing the instructor with support materials to facilitate student success. Some special features include the following:

- *Design Characteristics.* Each text-workbook in the LIFE Series, including *Calculator Math for Job and Personal Use,* features a larger typeface to make it easier for students to read.
- *Appropriate Content.* Real-life issues and skills are emphasized throughout the text, with relevant examples and illustrations provided.
- *Objectives.* Instructional objectives are clearly stated for each unit, letting students know what they will learn.

- *Checkpoints.* Checkpoints follow short segments of instruction and provide students with an opportunity to use what they have learned immediately.
- *Goals.* Goals are listed for each exercise to give students motivation and direction.
- *Study Breaks.* Each unit contains study breaks that provide a refreshing break from study and yet contribute to the global literacy goal of students.
- *Summaries.* A summary of the student's accomplishments appears at the end of each unit, providing encouragement and reinforcement.
- *Putting It Together.* The end-of-unit activities cover the theory presented in the Checkpoints and provide goals for students to measure their own skill development and success.
- *Glossary.* Important terms in the text are printed in bold and defined the first time they are used. These terms are listed and defined in the Glossary for easy reference.
- *Answers.* Answers for all the Checkpoints and Activities are provided at the back of the text-workbook and designed for easy reference to facilitate independent and self-paced learning.
- *Personal Progress Record.* Students keep track of their own progress by recording scores on a Personal Progress Record. Students can measure their own success by comparing their scores to evaluation guides provided for each unit. Whenever a student's total score for a unit is below the minimum requirement, the student may request a Bonus Exercise from the instructor.

## SPECIAL FEATURES OF THE INSTRUCTOR'S MANUAL

The instructor's manual provides general instructional strategies and specific teaching suggestions for *Calculator Math for Job and Personal Use,* along with supplementary bonus exercises and answers, testing materials, and a certificate of completion.

- *Bonus Exercises.* Second-chance exercises for all activities are offered through bonus exercises provided in the Instructor's Manual. These bonus exercises enable instructors to provide additional applications to students whose scores are less than desirable for a unit. Answers to all bonus exercises are also provided and can be duplicated for student use.

- *Testing Materials.* Four assessment tools, entitled "Checking What You Know," are provided. These tests may be used interchangeably as pre-tests or post-tests.
- *Certificate of Completion.* Upon completion of *Calculator Math for Job and Personal Use,* a student's success is recognized through a certificate of completion. This certificate contains a listing of topics that were covered in the calculator math program. A master certificate is included in the manual.

*Calculator Math for Job and Personal Use* is designed specifically to help you invest in your adult learners' futures and to meet your instructional needs.

# CONTENTS

# GETTING ACQUAINTED

Most of us can use a hand-held calculator to solve math problems. Almost every day, the amount of math used at work changes from job to job. However, jobs today do require that a worker solve math problems. A hand-held calculator can help solve these problems. You often need to use the calculator to solve math problems in the home, at work, and at school. You may need to use the calculator to figure how much you are saving when you use coupons at the grocery store, the number of hours you worked, or to add your grades in a course at school.

You may be enrolled in this math program because you know your math skills need improvement. Or you might be taking it to brush up on calculator skills. This book will help you to reach either objective.

## WHAT'S GOOD ABOUT CALCULATORS?

You will like working with a calculator whether you are good at math or if you need to improve your math. Everyone likes to work with calculators because it is easier, faster, and more accurate to press a key on a calculator than it is to add numbers in your head. While you are doing math by calculator, you will also be learning how to solve math problems.

## HOW YOU WILL LEARN

There is a system used in this book to help you learn. You need to know how that system works.

### Pre-Evaluation and Post-Evaluation

Your instructor has a pre-evaluation which may be given to you; it consists of calculator math problems. It checks the skill you have before you study from this book. A post-evaluation may also be used. This checks what you have learned though your study of calculator math. Your instructor will tell you what the standards are for the post-evaluation.

### Learn at Your Own Pace

You will go through the lessons in this book on your own. You can move at your own best pace. You may move ahead faster, or go slower, than other students. But don't be concerned about your pace. You should work at *your* best speed.

### Building on Success

Your study will be based on how well you learn each skill. This means that you are given goals called What You Will Learn. You will know what you are to achieve. You will study a topic and then check up on what you have learned. When you have shown that you know the topic, you move on to the next topic.

### Getting A Second Chance

You may not reach your assigned goal on every checkpoint and activity. When this happens, your instructor will ask you to review the lesson again and then do a Bonus Exercise, called a Bonus Checkpoint or Bonus Activity. These Bonus Exercises cover the same lessons as the regular exercises in the book. Your instructor has the complete set of Bonus Exercises. When you need one, ask for it. You score your own work, just as the regular exercises in the text. The practice exercises give you a second chance to reach your goal. When you score higher on a Bonus Checkpoint or Bonus Activity than you did on the original activity, you may change your score on your Personal Progress Record.

### Check Your Own Success

You will keep track of your own success. After each exercise in the book, you will check your own work. The answer key is at the back of the book. Then you will record your score on your own Personal Progress Record, also at the back of the book.

## WHAT YOU WILL LEARN

As you study this book you will learn the basic skills needed to do calculator math. The major topics are:

Get correct answers to addition, subtraction, multiplication, and division problems.

Write numbers so they are easy to read.

Proofread numbers, estimate answers, and check answers.

Solve word problems.

Work with fractions, decimals, and percents.

Use the calculator's memory.

Solve problems that use two or more of the arithmetic operations addition, subtraction, multiplication, and division.

Find averages.

Review what you have learned by completing car repair orders for a business.

## SPECIAL FEATURES

This book has a number of special features. These features will help you learn and apply math successfully.

### Break Times

Each unit in this book has break times. These give you a rest from study. The names of these break times are *Calculators Make It Easy; Just For Fun;* and *Yesterday, Today, and Tomorrow.* The Calculators Make It Easy sections show you how calculators can make math easier for you. The Just For Fun sections show you how to play games on the calculator. The Yesterday, Today, and Tomorrow sections tell you some of the history of calculators.

### Checkpoints

This book contains exercises called Checkpoints. Checkpoints will help you find out if you understand a topic before moving on to the next topic.

### Putting It Together

At the end of each unit is a section called Putting It Together. This section contains several activities similar to the Checkpoints within the units. They will help you apply and reinforce the skills you learned in the unit.

## Bonus Checkpoints and Activities

If you do not reach the assigned goal for any of the Checkpoints or Activities, you are asked to review the unit. Then you are asked to do a Bonus Checkpoint or Bonus Activity. These exercises give you a second chance to succeed. These second-chance activities are not in this book; your instructor has copies for you. Your instructor also has the answer key to these activities. You will use it to check your work.

## Answer Key

Answers to all of the checkpoints and activities are at the back of this book. The answer key pages are in color, making them easy to find and use. Use the key to check your work. Always do the exercises and activities *before* you look at the answers. Use the answers as a tool to check your work—not as a means of completing the work.

## Personal Progress Record

You will check most of your work against the answers. Then you will record your score on your Personal Progress Record at the back of the book. After you complete a unit, you will be able to judge the level of your success.

## Completion Certificate

When you finish your study in this book you may be eligible for a certificate of completion. Your instructor will explain the skill level required for this award.

## READY TO START!

You are now ready to start to use your calculator and improve your basic math skill for personal and job use. Throughout this book you will be given added directions. Through the guided study and the completion of the activities and projects in the book you will find that you will improve in your skill in doing calculator math.

If you have questions, your instructor or an aide will give you added materials and support.

# PART ONE
## BASIC MATH WITH YOUR CALCULATOR

### UNIT 1
ADDITION—THE BASICS

### UNIT 2
ADDITION—USING IT EVERY DAY

### UNIT 3
SUBTRACTION—USING IT EVERY DAY

### UNIT 4
MULTIPLICATION—USING IT EVERY DAY

### UNIT 5
DIVISION—USING IT EVERY DAY

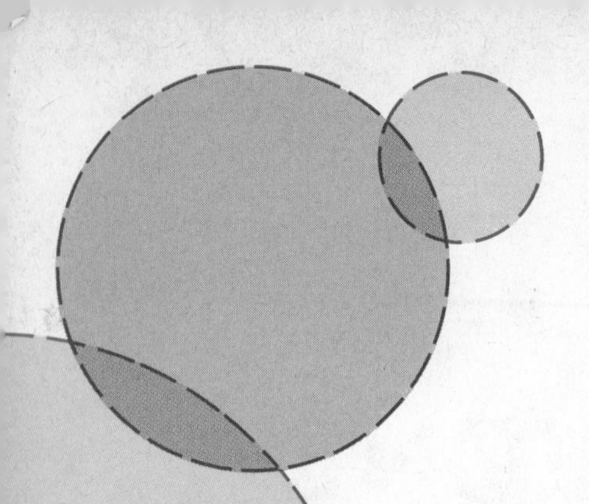
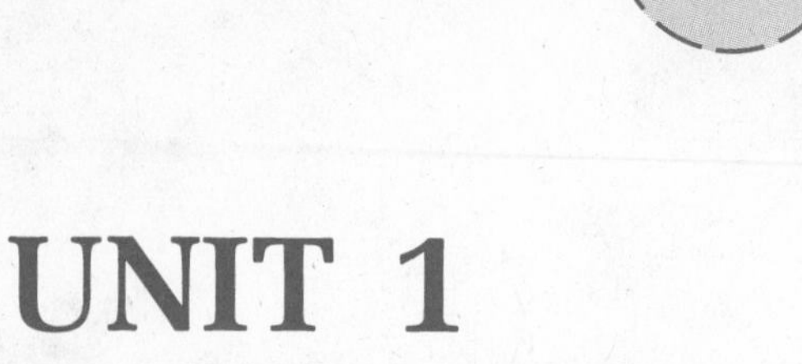

# UNIT 1
## ADDITION—THE BASICS

### WHAT YOU WILL LEARN

When you finish this unit, you will be able to:
- Get the correct answers to addition problems.
- Write the numbers 0–9 so they are easy to read.
- Proofread numbers with different styles and sizes.

## THE BASICS OF ADDITION

On the job and in your personal life, you will probably solve more addition problems than any other kind of math problem.

**Addition** means combining two or more numbers, called **addends,** to get a total. A **total** (also called a **sum**) is the answer to an addition problem. For example, if you were setting the supper table for three people in your family and suddenly two guests arrived, you would add two more place settings. The addends are 3 and 2. The total or sum is 5.

| | Place Settings | |
|---|---|---|
| Your family: | 3 | Addend |
| Guests: | + 2 | Addend |
| | 5 | Total (or Sum) |

Addends may have one or more digits. A **digit** is one of the numbers 0, 1, 2, 3, 4, 5, 6, 7, 8, or 9. When there are only a few addends with one digit each, it is easy to add the numbers in your head. For example, 3 + 2 = 5 or 2 + 4 + 7 = 13. When there are many numbers with many digits each, it is hard to add the numbers in your head. For example, 6,529 + 48 + 3,710. For this problem, it will be easier and more accurate to use a calculator.

Your boss at the Pastry Shop asks you to find out how many pastry boxes are in the storeroom. You count them and find there are 58 large, 39 medium, and 61 small boxes. You need to add the three addends to get the total. Since it is hard to add those numbers in your head, you will use a calculator.

## Calculator Addition

Your calculator may look like the one in Illustration 1-1.

Don't worry if your calculator is slightly different. It will still work the same as the one in the picture.

To add the addends 58, 39, and 61:

1. First, turn your calculator ON. Tap the [ON/C] On/Clear Key or [CE/C] Clear Entry/Clear Key. Some calculators are always ON and ready to use.
2. Clear any numbers out of the calculator. Tap the [AC] All Clear Key, or the [CE/C] Key.
3. Enter the addends by pressing these keys:

   58 + 39 + 61 = 158 _____

   The total should be in the answer window. Does your answer match the answer in the book (158)? If your answer is different from the book's, do the problem until you get the correct answer.
4. Write your answer on the blank line after the problem.
5. Clear the calculator again. Make it a habit to clear your calculator after each problem so the next problem starts with a 0 in the window.

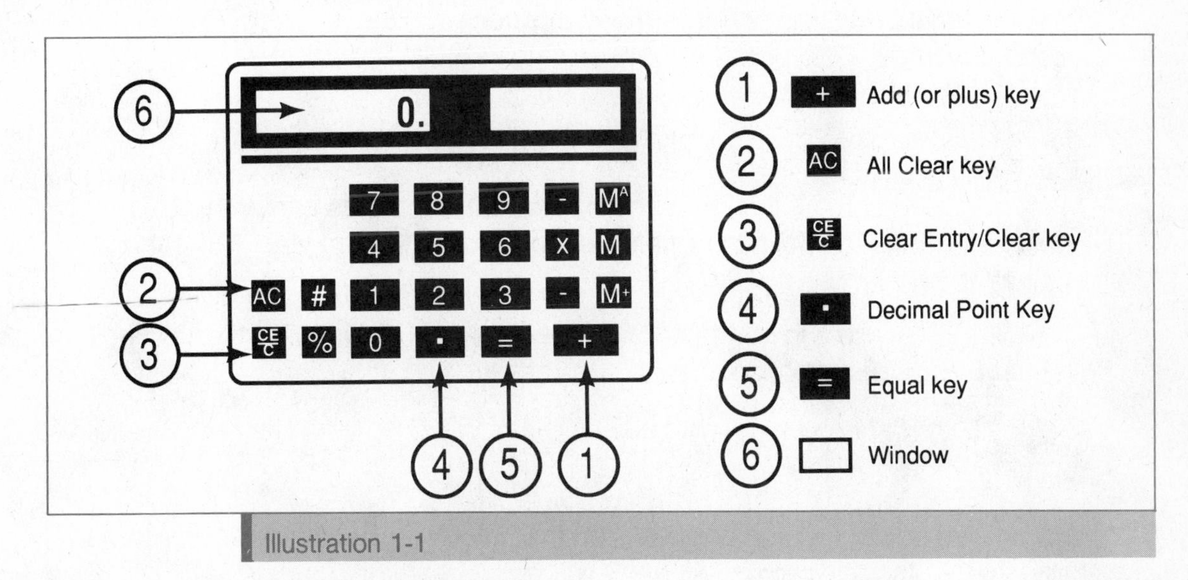


Illustration 1-1

A hand-held calculator.

## Vertical Addition

Sometimes addends will be listed **vertically** (up and down) instead of **horizontally** (from left to right). Enter the addends the same as you would for horizontal problems. For the following problem you would tap the keys 58 + 39 + 61 as you did before.

```
   58
   39
 + 61
```

## CHECKPOINT 1-1

**YOUR GOAL:** Get 4 of 5 answers correct.

Before you work these problems, find the number keys from 0 to 9, the [+] Plus Key, and the [=] Equal Key on your calculator. If your answers are not the same as the book's answers, do the problems until your answers are the same as the book's answers.

- ● 23 + 249 + 4 = 276
- **1.** 6 + 48 + 29 = ____
- **2.** 7 + 39 + 648 = ____
- **3.** 273 + 8 + 91 = ____
- **4.** 47 + 483 + 9 = ____
- **5.** 520 + 142 + 308 = ____

***Check your answers at the back of the book. Then count the number of answers you got correct. Record your score on the Personal Progress Record on page 141.***

### CALCULATORS MAKE IT EASY!

Most of us like work that is easy to do. Addition problems can be hard to do in your head when you have to remember a number and then "carry" it to the next column to the left.

```
   1
   58
   37
 + 61
  156
```

The 1 in 16 (the total of the right column) is mentally "carried" to the next column to the left.

You will like doing addition problems on the calculator because it will automatically "carry" the 1 over to the next column to the left.

## BEING ACCURATE

Calculators help you get totals quickly—but it is more important to be correct than to be fast. A total you get quickly is no good if it is not correct. Here is what you can do to get correct totals:

## See, Read, Remember, and Write Totals

To get a correct answer, you must take time to see and read each number correctly, remember the number in your head, and then enter that same number into the calculator.

Think about what you are doing. Read numbers in groups if they are separated by commas. Pause where there is a comma as in Illustration 1-2.

Illustration 1-2

See, read, remember, and write numbers.

| See, Read | Remember |
|---|---|
| a. 89 | eight nine |
| b. 574 | five seven four |
| c. 6,458 | six [pause] four five eight |

## CHECKPOINT 1-2

**YOUR GOAL:** Get all answers correct.

Write in the blank spaces what you see, read, and remember.

| | See and Read | Write |
|---|---|---|
| ● | 54 | five four |
| **1.** | 72 | __________ |
| **2.** | 371 | __________ |
| **3.** | 698 | __________ |
| **4.** | 8,964 | __________ |
| **5.** | 2,514 | __________ |

☞ ***Check your work. Record your score.***

## Write Numbers Legibly

Writing numbers **legibly** means writing them clearly. You must write numbers legibly so that you and others can read them quickly and correctly. To avoid mistakes, take the time to form your numbers carefully.

When numbers are not written legibly, it is hard to know what the numbers are. For example, look at Illustration 1-3. Write the

Illustration 1-3

Illegible handwritten numbers.

| | Column A | Column B |
|---|---|---|
| | 987 | |
| | 083 | |
| | 223 | |
| | 690 | |
| | 488 | |
| | 230 | |
| | 945 | |
| Total | 3874 | |

numbers that you believe are represented in Column A in Column B. Do your numbers add up to the same total?

## CHECKPOINT 1-3

**YOUR GOAL:** Get a 3 rating for Step 3. The ratings are:

* = 1
** = 2
*** = 3
**** = 4

To help you write numbers legibly, do this task:

*Step 1*: Write the numbers 0 through 9 on the line below:

________________________________________

*Step 2*: Now compare your numbers with those in Illustration 1-4 on the 4-star (****) line of the Number Rating Scale.

NUMBER RATING SCALE

| *Rating* | *Example* | *Descriptio* |
|---|---|---|
| **** | 0123456789 | Perfect legibility |
| *** | 0123456789 | Legible |
| ** | 0123456789 | Minimum legibility |
| * | 0123456789 | Almost illegible |
| 0 | 0123456789 | Illegible |

Illustration 1-4

Numbers rated from Perfect Legibility to Illegible.

(1) Are all the numbers the same size?
(2) Are the lines straight in the numbers 1, 4, and 7?
(3) Are the angles in the numbers 4, 5, and 7 straight and not rounded?
(4) Are the circles in the numbers 6, 8, 9, and 0 closed?
(5) Are the sizes of the numbers and the spaces between numbers written so they fit the length of the line on which they are written?

*Step 3*: Writing legible numbers requires attention and careful practice. Practice again writing legibly the numbers 0 through 9 until they have a 3-star or 4-star rating.

---

☞ ***Check your work. Record your score.***

## Prove the Total

To **prove the total** for an addition problem means to do the problem a second time in a different way. The second time, add from the bottom to the top of the problem. The chance of you making the same mistake a second time is less if you add them in a different order. For example, if you first add the numbers from bottom to top, next time add from top to bottom. Or from the right to the left instead of from the left to the right. To prove a total:

*Step 1*: Clear your calculator.

*Step 2*: Work the problem by adding down:

*Add Down*

$$\begin{array}{r} 58 \\ 39 \\ +\ 61 \\ \hline \end{array} \downarrow$$

Record the answer in the blank space.

*Step 3*: Prove the first answer by adding up. If the second answer is:

(a) the *same* as the first answer, you can assume the answer is correct.
(b) *different* from the first answer, repeat the problem until you get two answers that match. Draw a line through the incorrect answer.

*Add*
*Up*

```
  58  ↑
  39  |
+ 61  |
```

Record the answer in the blank space.

*Step 4*: For horizontal problems, add from left to right:

→
58 + 39 + 61 = ____

and then from right to left to prove the total:

←
58 + 39 + 61 = ____

## Proofread Numbers

**Proofreading numbers** means examining (looking at) digits in a number to see if they are the same or different. Numbers may appear in different styles and sizes, as shown in Illustration 1-5.

Illustration 1-5

Proofread carefully.

| Column A | Column B |
| --- | --- |
| 1. 04371 | 04317 |
| 2. 32,413.88 | $2,43188 |
| 3. 471983 | 471938 |
| 4. May 13, 1987 | May 13, 1987 |
| 5. Pages 324-491 | Pages 342-419 |

## CHECKPOINT 1-4

**YOUR GOAL:** Get 4 of 5 answers correct for Column C only.

Draw a line through any number in Column B that is different from the number to be matched in Column A. Write the correct number in Column C. If the numbers in Columns A and B are the same, write the number in Column C. In Column D, rate your writing with stars after comparing each item with the examples in the Number Rating Scale, on page 6.

| | *Column A* | *Column B* | *Column C* | *Column D* |
| --- | --- | --- | --- | --- |
| ● | 72HB609 | ~~72HB909~~ | 72HB609 | * * * * |
| **1.** | 6431 ÷ 839 | 6431 ÷ 839 | ____ | ____ |
| **2.** | 11:03 a.m. | 11:03 a.m. | ____ | ____ |

**3.** 214-401-5970 214-481-4970 ____________ ____________

**4.** 8/14/88 8/14/88 ____________ ____________

**5.** 368 907 368 907 ____________ ____________

☞ ***Check your work. Record your score.***

## WHAT YOU HAVE LEARNED

Now that you have completed this unit, you should be able to:

- Correctly add numbers.
- Write numbers legibly.
- Proofread typewritten and handwritten numbers.

# PUTTING IT TOGETHER

## ACTIVITY 1-1 YOUR GOAL: Get all answers correct.

Write your answers in the blank spaces.

What is the total?

| | | | | | |
|---|---|---|---|---|---|
| ● | 274 + | 6 + | 19 | = | 299 |
| **1.** | 649 + 382 + | 27 + | 570 | = | ______ |
| **2.** | 728 + 385 + | 482 + | 42 | = | ______ |
| **3.** | 2,819 + 349 + | 5,791 | | = | ______ |
| **4.** | 315 + 827 + | 98,630 | | = | ______ |
| **5.** | 128 + 46 + | 4,318 + | 24,891 | = | ______ |

***Check your work. Record your score.***

## ACTIVITY 1-2 YOUR GOAL: Get all answers correct.

The correct numbers are in Column A. Draw a line through any number in Column B that is different from the number to be matched in Column A. Write in Column C the correct numbers from Column A.

| | Column A | Column B | Column C |
|---|---|---|---|
| ● | 61897 | ~~61391~~ | 61897 |
| **1.** | **$1,892.43** | $1,982.43 | ______ |
| **2.** | 12:38 p.m. | 12:08 p.m. | ______ |
| **3.** | SQ-178-3062 | SO-718-3062 | ______ |
| **4.** | 9/24/93 | 9/24/96 | ______ |
| **5.** | 15973628 | 160180ZS | ______ |

***Check your work. Record your score.***

# UNIT 2

## ADDITION—USING IT EVERY DAY

### WHAT YOU WILL LEARN

When you finish this unit, you will be able to:

- Round numbers to tens, hundreds, and thousands places.
- Estimate the answers to addition problems.
- Use a six-step process to solve word problems.
- Get the correct answers to addition, decimal, and money problems.

## ROUNDING NUMBERS

To **round** a number means to change a number to the nearest power of ten such as tens (10), hundreds (100), or thousands (1,000). Rounded numbers make it easier to quickly understand amounts. For example, it is easier to "grasp in your mind" $10 instead of $9.95. Each digit in a number has a value depending on its place in a number. Illustration 2-1 shows the place value of digits in the number 7,146.

Illustration 2-1

Place value chart.

Places

| Thousands | Hundreds | Tens | Ones |
|---|---|---|---|
| 7 | 1 | 4 | 6 |

The 6 is in the *ones* place.
The 4 is in the *tens* place.
The 1 is in the *hundreds* place.
The 7 is in the *thousands* place.

To round 7,146 to the nearest *hundred*:

1. Look at the digit in the place you are rounding and underline it. (Underline 1 because 1 is in the *hundreds* place and we are rounding to the nearest hundred.)
2. Look at the digit to the right of the place you are rounding and circle it. (Circle 4 because it is to the right of 1.)
3. If the digit is
   a. 5 or *more,* add 1 to the place you are rounding. (4 is *not more than 5,* so go to step b.)
   b. *less* than 5, keep the same number in the place you are rounding. (4 is *less than 5,* so keep the 1.)
4. Change all digits to the right of the place you are rounding (1) to zeros. (Change the 4 to 0 and the 6 to 0 so 146 changes to 100. 7,146 is rounded to 7,100, the nearest hundred.)

Now round 7,146 to the nearest *ten* by following the steps below and Illustration 2-2:

1. The rounding number is 4 in the *tens* place. Underline it.
2. The number to the right of 4 is 6. Circle it.
3. 6 is more than 5. So add 1 to the 4, the place you are rounding, which changes the 4 to 5.
4. Change the 6 to 0. 7,146 rounded to the nearest ten becomes 7,150.

Illustration 2-2

Rounding steps.

*Step*

1. 7,146
2. 7,146
3. 7,15__
4. 7,150

## CHECKPOINT 2-1

**YOUR GOAL:** Get 4 of 5 answers correct.

Round each number to the place shown.

● 2,476 to the nearest tens place. **2,480**

**1.** 326 to the nearest tens place. ______

**2.** 5,078 to the nearest hundreds place. ________

**3.** 3,479 to the nearest thousands place. ________

**4.** 7,551 to the nearest hundreds place. ________

**5.** 4,623 to the nearest thousands place. ________

***Check your work. Record your score.***

## ESTIMATING TOTALS

"Accuracy first" is our motto for all math problems. You already know some ways to get the correct answer from the Being Accurate section. Another accuracy check is to **estimate** (or judge) the total.

To estimate the total of 47 + 631:

1. Round each addend to the highest place at the left. (47 becomes 50 and 631 becomes 600.) Write the rounded addends in the Rounded Addends column.
2. Mentally add the Rounded Addends and write the total 650 in the Estimated Total column.
3. Use the calculator to add the original addends. (47 + 631). Write the total in the Calculator Total column (678).
4. If the Calculator Total is not close to the Estimated Total, estimate and calculate the answer again. Write the correct Estimated Total and Calculator Total.

| Original Addends | Rounded Addends | Estimated Total | Calculator Total |
|---|---|---|---|
| 47 + 631 | 50 + 600 | 650 | 678 |

## CHECKPOINT 2-2

**YOUR GOAL:** Get 12 of 15 answers correct.

Work the following problems.

| | Original Addends | Rounded Addends | Estimated Total | Calculator Total |
|---|---|---|---|---|
| ● | 691 + 178 | 700 + **200** | 900 | 869 |
| **1.** | 17 + 41 | ________ | ________ | ________ |
| **2.** | 34 + 740 | ________ | ________ | ________ |
| **3.** | 271 + 426 | ________ | ________ | ________ |

**4.** 1,729 + 5,493 ________ ________ ________

**5.** 1,830 + 46,925 + 7,026 ________ ________ ________

☞ ***Check your work. Record your score.***

## SOLVING WORD PROBLEMS—ADDITION

In textbooks, math problems are often written as number problems (with no words) such as:

$$\begin{array}{r} 56 \\ 39 \\ +\ 61 \\ \hline \end{array}$$

In real life, math problems are not "set up" for you but are part of a "story" that you must study to get the answer. These problems are sometimes called **word problems.** For example:

Your boss asks you to find the *total* number of boxes in the storeroom. You count the boxes and find out that there are 58 large, 39 medium, and 61 small boxes.

### Problem-Solving Process

There is an easy six-step process that will help you solve math word problems.

| Step | To Do | Explanation | Example |
|---|---|---|---|
| 1 | Read the story | Look for and underline key words. | *total number of boxes, 58 large, 39 medium, 61 small* |
| 2 | Answer key questions | What information is:<br>a. Needed?<br>b. In the story? | <br>a. *Total number of boxes*<br>b. *58, 39, 61* |
| 3 | Decide which math operation to use | +, −, ×, or ÷ | The story says to "total" so you must add. |
| 4 | Estimate the answer | About what should the answer be? | The answer will be about 150 since all of the addends are about 50 and mentally you know that three 50's = 150. |
| 5 | Work the problem | Calculate accurately. | 58 + 39 + 61 = 158 |
| 6 | Prove the answer | Calculate the addends in reverse order. | 61 + 39 + 58 = 158 |

## Clues to Look For

You should use addition when words in a problem mean the same as *add.* For example, in the following story "put together" means the same as *add,* so this is an addition problem.

James and Jenny want to buy a Mother's Day gift. James has $6 and Jenny has $8. How much can they spend on a gift if their money is put together?

Some examples of words that mean the same as *add* are:

accumulate
all together
combine or combines or combined
in all
join or joins or joined
put together
sum
total

These words (and other words like them) are the clues that will tell you if the story is an addition problem.

## CHECKPOINT 2-3

**YOUR GOAL:** Get 4 of 5 answers correct.

Read the following stories and find the "clues" that will tell you if it is an addition problem. Underline the clue.

If it is an addition problem, get the answer. If it is not an addition problem, write "No +" in the answer space.

- You work at Bigger Burgers. The supervisor asks you to take inventory of the cups in the stockroom and front serving area. You count 165 cups in the stockroom and 59 cups in the front serving area. How many cups are there in all?

$165 + 59 =$ 224

1. Pauline wants to get a hair dryer at the trading stamp store with her stamps. Her stamp books are in three stacks; 8 in one stack, 9 in another stack, and 16 in another stack. How many books does she have all together?

________

2. Mary has $15 in her purse. She finds a pair of shoes on sale for $10 and decides to buy them. How much money does she have left in her purse?

________

3. Carl and Sylvia decided to shop for a new car. They saw one they liked. A sheet of paper taped to the car window showed the price of the car and other charges:

| | |
|---|---|
| Base Price | $5,995 |
| AM/FM Stereo Radio | 235 |
| Automatic Transmission | 180 |
| Air Conditioning | 574 |
| Destination Charge | 246 |

What is the total price of the car? ______

**4.** George and Fred were hired to paint a garage for Ms. Bush. Ms. Bush paid them $200. How much did George make after the money was divided evenly between the two men?

______

**5.** Donna was shopping at a second-hand store. She decided to buy a purse for $8, a pair of shoes for $13, and a pair of sandals for $9. How much did she spend at the store?

______

☞ ***Check your work. Record your score.***

## DECIMALS AND AMOUNTS OF MONEY IN ADDITION

The [·] Decimal Point Key is pressed at the place in the addend where the decimal occurs. To enter 3.81 + 0.75 in the calculator, press 3 [·] 81 + [·] 75 = . The total is 4.56.

Amounts of money are written as dollars and cents with two decimal places. To get the total of a grocery purchase, tap the numbers and decimal keys as shown below:

```
  $ 3.81
    2.25
    1.02
+    .79
  ------
  $ 7.87   ______
```

Record your answer on the blank line.

## DECIMALS NOT ALIGNED

Sometimes decimals will not be **aligned** (stacked in a straight line below each other). Just enter the decimal point at its proper place in each addend. The answer will appear in the window with the decimal accurately placed, as shown in Illustration 2-3.

```
    1.736
    5.98
+   2.4
```

Illustration 2-3

Your calculator automatically aligns the decimal.

## CHECKPOINT 2-4

**YOUR GOAL:** Get 4 of 5 answers correct.

Find the totals.

```
●      21
       3.78
  + 475.95
    500.73
```

```
1.   15.80    2.  $42.39    3.  $ 92.37    4.   5.633    5.    .67
     23.67          3.67           8.98          7.89          43.27
  +   4.05     +   55.55     +   639.24     + 138.4       + 1.7849
```

*Check your work. Record your score.*

## CALCULATORS MAKE IT EASY!

We like to use calculators—they make our work easier than doing the same work by hand or in our head. For example, to add numbers with decimals without using a calculator, take the time to rewrite all of the addends with unaligned decimals in Column A so the decimals will be aligned in Column B:

| *Column A* | *Column B* |
|---|---|
| Unaligned decimals | Addends rewritten to align decimals |
| 6.049 | |
| 26.1 | |
| 7.37 | |
| + 857.3 | ______ |
| | ______ |

Now, mentally add the numbers you just rewrote in Column B and record the answer.

To compare how easy it is to use a calculator for addends with decimals, add the numbers in Column A with your calculator.

## WHAT YOU HAVE LEARNED

Now that you have completed this unit, you should be able to:

- Round numbers.
- Estimate answers for addition problems.
- Correctly add numbers with decimals and amounts of money.
- Solve word problems.

# PUTTING IT TOGETHER

## ACTIVITY 2-1 YOUR GOAL: Get 4 of 5 answers correct.

Round each number to the place suggested.

- 5,291 to the nearest thousands place. **5,000**

1. 29 to the nearest tens place. ______
2. 83 to the nearest tens place. ______
3. 758 to the nearest hundreds place. ______
4. 3,649 to the nearest thousands place. ______
5. 9,148 to the nearest thousands place. ______

☞ ***Check your work. Record your score.***

## ACTIVITY 2-2 YOUR GOAL: Get 12 of 15 answers correct.

Work the following problems.

| | Original Addends | Rounded Addends | Estimated Total | Calculator Total |
|---|---|---|---|---|
| ● | 871 + 392 | 900 + **400** | 1300 | 1263 |
| 1. | 35 + 54 | | | |
| 2. | 47 + 197 | | | |
| 3. | 639 + 216 | | | |
| 4. | 799 + 924 | | | |
| 5. | 108 + 593 | | | |

☞ ***Check your work. Record your score.***

## ACTIVITY 2-3 YOUR GOAL: Get 4 of 5 answers correct.

Where necessary, underline the clues to these addition problems. Then work the problems to find the correct totals.

- Jamie is in a weight-reducing class. In four months she lost 14 pounds, 12 pounds, 7 pounds, and 4 pounds. How many pounds did she lose all together?

**37**

1. Your family drove to another state to visit relatives. The first day you traveled 328 miles, the second day 297 miles, and the last day 405 miles. In all, how many miles did your family drive?

_________

2. Rob wants to estimate the miles to his aunt's house. The map says the distance is 383 miles. Round these miles to hundreds.

_________

3. Janet is planning to buy a pair of shoes for about $39, a blouse for about $22, and a pair of pants for about $38. About how much money will she need all together?

_________

4. Kaye worked in a grocery store 37 hours the first week of August, 29 hours the second week, 40 hours the third week, and 33 hours the fourth week. How many hours did she work total?

_________

5. Mandy went to the grocery store with very little money. Before checking out, she wants to make sure she has not put too many groceries into her cart. Below is a list of prices of the items she had in her grocery cart. What was the total of the grocery items?

$$\begin{array}{r} \$0.39 \\ 0.69 \\ 1.19 \\ 0.89 \\ +\ 0.24 \\ \hline \\ \hline \end{array}$$

☞ ***Check your work. Record your score.***

## ACTIVITY 2-4 YOUR GOAL: Get 4 of 5 answers correct.

Write the totals in the blank spaces.

● 
$$\begin{array}{r} 58.53 \\ 2.31 \\ .20 \\ 9.32 \\ +\ 25.80 \\ \hline \mathbf{96.16} \end{array}$$

| 1. | 2. | 3. | 4. | 5. |
|---|---|---|---|---|
| 25.63 | \$36.45 | \$19.28 | 793.3 | 9.263 |
| .78 | 2.98 | 26.43 | 2.86 | 43.8 |
| 9.41 | .37 | .29 | 5.941 | .57 |
| 43.07 | .03 | 1.56 | .3709 | .453 |
| + .56 | + 19.85 | + .75 | + 520.8 | + 673.1 |

☞ ***Check your work. Record your score.***

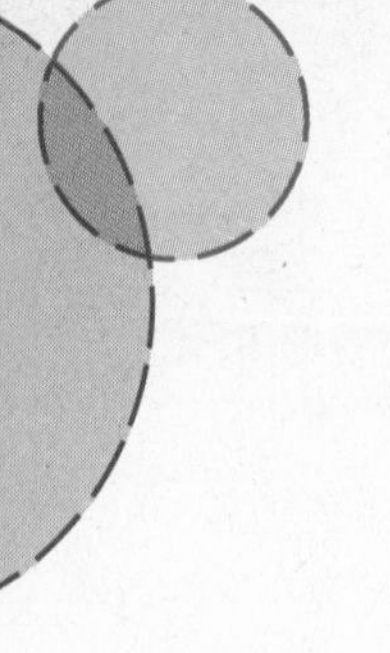
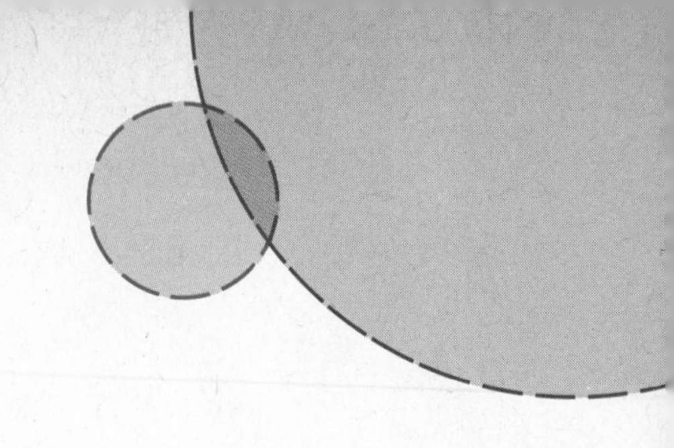

# UNIT 3

## SUBTRACTION—USING IT EVERY DAY

### WHAT YOU WILL LEARN

When you finish this unit, you will be able to:

- Get the correct answers to subtraction problems.
- Prove the answers to subtraction problems.
- Estimate the answers to subtraction problems.
- Find clues to subtraction problems.
- Get the correct answers to subtraction of decimals, amounts of money, and series of subtrahends problems.

## THE BASICS OF SUBTRACTION

**Subtraction** is taking one number away from another number. The number *being subtracted* (the lower number) is called the **subtrahend,** and the number *being subtracted from* (the upper number) is called the **minuend.** The answer to a subtraction problem is called the **difference.**

```
   59  Minuend
 - 46  Subtrahend
 ----
   13  Difference
```

### Calculator Subtraction

Suppose the starter on your car does not work. You need to buy a new one that costs $43 so you can get to work. You now have only $16. How much more money do you need to buy the starter? You need to subtract the amount of money you have from the price of the starter to find how much more money you need. This is difficult to do in your head, so you will use a calculator.

To subtract 16 from 43:

1. Make sure your calculator is on.
2. Clear the calculator by pressing the [AC] key, or the [CE/C] key.

3. Enter the minuend, 43.
4. Locate the [–] Minus Key on your calculator as shown in Illustration 1-1 on page 000. Press the [–] Minus Key.
5. Enter the subtrahend, 16.
6. Press the [=] Equal Key.

The answer, 27, should be in the answer window. If it is not, work the problem until you get the correct answer.

## Horizontal Subtraction

Minuends and subtrahends can be written vertically or horizontally, such as: 193 – 52 = 141. For this problem, press the keys **193 – 52 =**. Subtraction problems written horizontally are worked the same way as problems written vertically.

## PROVE THE DIFFERENCE

To prove answers in subtraction problems, add the difference to the subtrahend. If the total is equal to the minuend, your answer is correct. In other words, difference + subtrahend = minuend. For example:

| | | | | |
|---|---|---|---|---|
| 59 | Minuend | Proof: | 46 | Difference |
| – 13 | Subtrahend | | + 13 | Subtrahend |
| 46 | Difference | | 59 | Minuend |

## CHECKPOINT 3-1

**YOUR GOAL:** Get all answers correct.

To practice solving subtraction problems listed vertically and horizontally, work the problems below. Prove your answers by adding the difference and the subtrahend to get the minuend. Write your answers in the blank spaces.

| | | | | |
|---|---|---|---|---|
| ● 87 | **Proof:** | 56 | Difference |
| – 31 | | + 31 | Subtrahend |
| 56 | | 87 | Minuend |

**1.** 35 – 14 = ____

**2.** 79 – 34 = ____

**3.** 343 – 89 = ____

**4.** 562 – 231 = ______

**5.** 7,430 – 856 = ______

☞ ***Check your work. Record your score.***

## YESTERDAY, TODAY, AND TOMORROW

The first electronic calculator was introduced about 1970. It cost about $1,200. It was as large as a typewriter. One of the latest calculators, like yours, is about the size of a business card. Some small calculators cost about $4 to $10 and calculate faster than the early $1,200 calculators.

1980 Model

1990's Model

## ESTIMATING DIFFERENCES

We estimate the difference of a subtraction problem to help us find errors. Estimating the answer of a subtraction problem is similar to the process of estimating the answer of an addition problem.

To estimate the difference for the problem 519 − 98:

1. In your head, round the minuend and the subtrahend to the highest place at the left. (519 becomes 500 and 98 becomes 100.) Write the rounded numbers in the Rounded Problem column.
2. Mentally subtract the rounded subtrahend from the rounded minuend and write the difference (400) in the Estimated Difference column.
3. Use the calculator to work the original problem (519 − 98). Write the difference in the Calculator Difference column (421).
4. If the Calculator Difference is not close to the Estimated Difference, estimate and calculate the difference again. Write the correct Estimated Difference and Calculator Difference.

| *Original Problem* | *Rounded Problem* | *Estimated Difference* | *Calculator Difference* |
|---|---|---|---|
| 519 − 98 | 500 − 100 | 400 | 421 |

## CHECKPOINT 3-2

**YOUR GOAL:** Get 12 of 15 answers correct.

Estimate and calculate the differences to these problems and write your answers in the blank spaces.

| | Original Problem | Rounded Problem | Estimated Difference | Calculator Difference |
|---|---|---|---|---|
| ● | 52 − 18 | 50 − 20 | 30 | 34 |
| **1.** | 83 − 44 | ______ | ______ | ______ |
| **2.** | 865 − 79 | ______ | ______ | ______ |
| **3.** | 184 − 92 | ______ | ______ | ______ |
| **4.** | 370 − 165 | ______ | ______ | ______ |
| **5.** | 276 − 138 | ______ | ______ | ______ |

☞ ***Check your work. Record your score.***

## SOLVING WORD PROBLEMS—SUBTRACTION

You should use subtraction when the words in a problem mean the same as *subtract.* Some examples of words that mean the same as *subtract* are:

decrease or decreases or decreased
difference
left or left over
less or lessened
minus
reduce or reduces or reduced
remain or remains or remained
take away or took away

These words plus others like them are clues that let you know when to subtract. For example, *left over* is the clue in the following story that tells you to subtract.

Kristy was sick. The doctor gave her a bottle with 21 pills in it to cure her cold. The doctor said to take three pills a day for seven days. After five days, she had taken 15 pills. How many pills were left over in the bottle?

**6**

## CHECKPOINT 3-3

**YOUR GOAL:** Get 4 of 5 answers correct.

Read the following stories and underline the clues that tell you it is a subtraction problem. Work the problems and write your answers in the blank spaces.

- In Julio's night class at the Adult Learning Center, there were 17 students the first day. Julio thought it would be great if he talked to some of his friends and they decided to come. About three days later, Julio counted 23 students in his class. What is the difference in the number of students on the first day of class and three days later?

  6

1. Jack needed to buy a used car. He found a car that got 27 miles to the gallon in town. He compared this to his friend's car that gets 19 miles to the gallon in town. What is the difference in miles per gallon between the two cars?

   ______

2. David weighed 220 pounds. He wants to lose weight so he will feel and look better. He went on a diet and now weighs 26 pounds less. How much does he weigh now?

   ______

3. Cindy went to the store to buy a can of corn for 69 cents. She has a coupon that reduces the price by 13 cents. How much will she pay for the can of corn?

   ______

4. You borrowed $26 to pay off an electric bill. You were soon able to pay back $17. How much money remains to be paid back?

   ______

5. Chelsea decided to make brownies for her son's birthday party. She made 24 brownies. The children at the party ate 17 of the brownies. How many were left for her family to munch on?

   ______

☞ ***Check your work. Record your score.***

## DECIMALS AND AMOUNTS OF MONEY

Decimals in subtraction are used the same way as decimals in addition. Sometimes the decimal points may not be aligned in a problem. Enter the decimal point at its proper place in each number. The answer will be in the window with the decimal in the right place.

## CHECKPOINT 3-4

**YOUR GOAL:** Get 4 of 5 answers correct.

Find the differences and write them in the blank spaces.

| | |
|---|---|
| ● | $345.92 |
| − | 68.10 |
| | $277.82 |

| **1.** | | **2.** | | **3.** | | **4.** | | **5.** | |
|---|---|---|---|---|---|---|---|---|---|
| | $59.95 | | 25.638 | | 10.46 | | 1634.7 | | 193.65 |
| − | 3.65 | − | .5326 | − | .299 | − | 1.329 | − | 49.47 |

***Check your work. Record your score.***

As with addition, amounts of money are written as dollars and cents with two decimal places. One number in a money problem may have a decimal ($39.64) and another may not, ($23). Enter numbers and decimals exactly as they are read. For example: **$39.64 − $23** = ______. Press the [=] Equal Key. The calculator will then place the decimal in its correct place in the difference. Record your answer in the blank space. The difference should be $16.64.

You may need to make subtractions when using coupons at a grocery store. This is easy to do. Just press the [−] Minus Key and the next subtrahend. Continue to do this until you have entered all of the subtrahends. Then press the [=] Equal Key. Enter the minuend and the subtrahends:

$31.28 − $.25 − $.75 − $.10 = ______

Record your answer in the blank space. Did you get $30.18?

The series of subtrahends may be listed vertically:

| | |
|---|---|
| | 72.23 |
| − | .432 |
| − | 69.1 |
| − | .29 |

The difference in the window of your calculator should be 2.408. Use the same process for subtracting several subtrahends in money problems. Here is an example:

$102.06 − $53.72 − $34.19 − $7.48 = ______

The difference in this problem should be $6.67.

## CHECKPOINT 3-5

**YOUR GOAL:** Get 4 of 5 answers correct.

Work these problems and write your answers in the blank spaces.

● $22.57 − $.15 − $.33 − $.25 = **$21.84**

**1.** 10.6 − 5.23 = ______

**2.** $98.46 − 73.51 = ______

**3.** 10.95 − 1.489 − .352 = ______

**4.** $19.28 − $7.46 − $.57 − $.69 = ______

**5.** 54.7 − .384 − 1.539 = ______

☞ ***Check your work. Record your score.***

### JUST FOR FUN

Find the answers to the following questions with your calculator. Work the problem and then turn your calculator upside down to read the answer.

1. What does a fish breathe with?

   65,218 − 4,930 − 1,929 − 643 = ______

2. Sometimes you get one in your cake batter.

   88,345,776 − 11,000,113 = ______

3. What is a round ball that shows all of the land and water on the earth?

   1,258,746 − 1,220,670 = ______

4. A snake's voice sounds like this.

   5,612 − 98 = ______

5. What do babies wear when they eat?

   1,120 − 302 = ______

## WHAT YOU HAVE LEARNED

By completing this unit, you can now:

- Subtract numbers on the calculator with 100 percent accuracy.

- Prove differences.
- Estimate subtraction problem answers.
- Find clues that let you know when a word problem is a subtraction problem.
- Subtract decimals, amounts of money, and series of subtrahends.

## ACTIVITY 3-1 **YOUR GOAL:** Get all answers correct.

Write your answers in the blank spaces. Prove your answers. What is the total?

● $\begin{array}{r} 250 \\ -\ 36 \\ \hline 214 \end{array}$

1. $\begin{array}{r} 89 \\ -\ 54 \\ \hline \end{array}$

2. 455 − 96 = ______

3. 148 − 73 = ______

4. $\begin{array}{r} 675 \\ -\ 36 \\ \hline \end{array}$

5. $\begin{array}{r} 3{,}576 \\ -\ 245 \\ \hline \end{array}$

☞ ***Check your work. Record your score.***

## ACTIVITY 3-2 **YOUR GOAL:** Get 12 of 15 answers correct.

Estimate and calculate the differences for these problems and write your answers in the blank spaces.

| | Original Problem | Rounded Problem | Estimated Difference | Calculator Difference |
|---|---|---|---|---|
| ● | 56 − 19 | 60 − 20 | 40 | 37 |
| 1. | 52 − 21 | ______ | ______ | ______ |
| 2. | 236 − 49 | ______ | ______ | ______ |
| 3. | 464 − 92 | ______ | ______ | ______ |
| 4. | 179 − 57 | ______ | ______ | ______ |
| 5. | 382 − 264 | ______ | ______ | ______ |

☞ ***Check your work. Record your score.***

## ACTIVITY 3-3 **YOUR GOAL:** Get 4 of 5 answers correct.

Underline the clue that tells you to subtract in a word problem. Find the difference.

- The total of Miss Johnson's grocery bill is $13.92. She gave the cashier $15.00. How much money does she have left?

  **$1.08**

1. Jacob's rent for his apartment increased from $189 to $245. What is the difference between the two amounts?

   ______

2. Cindy bought a head of lettuce for $0.73. When she handed the cashier a $1 bill, how much change did she get back?

   ______

3. Andrea had $200 in her budget for a new washing machine. Her old one won't work and can't be repaired. She found a machine at a garage sale for $125. How much less did she spend than she thought she would?

   ______

4. Mary's grocery bill total came to $12.36. She gave the cashier $15. How much money does she have left over?

   ______

5. On October 8, Annie Moore's balance in her savings account was $192.31. She withdrew $23. How much money remains in her account?

   ______

***Check your work. Record your score.***

## ACTIVITY 3-4 **YOUR GOAL:** Get 4 of 5 answers correct.

Find the differences and write your answers in the blank spaces.

- 4.75 − .39 = **4.36**

| 1. | 2. | 3. | 4. | 5. |
|---|---|---|---|---|
| 26.931 | $357.45 | 59,386.1 | $94.75 | 360.4 |
| − 7.48 | − 18.90 | − .427 | − 2.81 | − 19.28 |
| | | | − 30.97 | − .745 |

***Check your work. Record your score.***

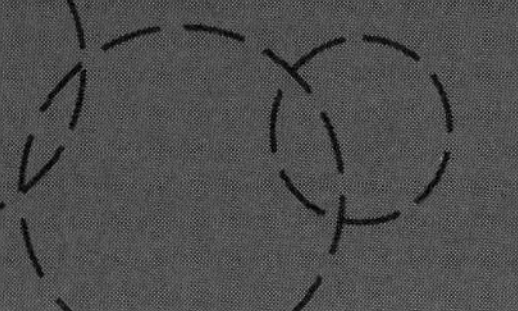

# MAKING IT WORK

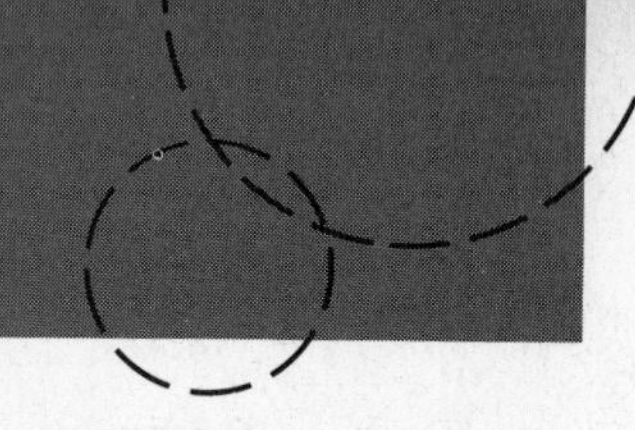

**REVIEW 3-1** **YOUR GOAL:** Get 5 of 7 answers correct.

## Clothing on Sale

Mr. and Mrs. Mattison went to a summer sale at Wearwell's clothing store to buy clothes for their children. They want to find out how much they saved by going to the sale. Below is a list of items that they bought, the sale price, and the regular price. Follow these steps to find out how much they saved on each item.

1. Subtract the Regular Price from the Sale Price and write the answer in the Difference column.

2. Repeat step 1 for the other items.

| | Regular Price | Sale Price | Difference |
|---|---|---|---|
| Jacket | $42.38 | $34.98 | ______ |
| Pants | $37.99 | $19.95 | ______ |
| Socks | $7.00 | $4.19 | ______ |
| Shirt | $18.99 | $14.25 | ______ |
| Sweater | $27.99 | $22.49 | ______ |
| TOTALS | ______ | ______ | ______ |

Find out how much they saved all together.

1. Add the Regular Price column and write the total in the blank space.

2. Add the Sale Price column and write the total in the blank space.

3. Subtract the Sale Price column from the Regular Price column and write the difference in the blank space.

☞ **Check your work by adding the Difference column. Record your score.**

## REVIEW 3-2 **YOUR GOAL:** Get 6 of 7 answers correct.

### Coupons in Supermarket

Brenda went to the supermarket to buy groceries with coupons she cut out of the newspaper. Follow the steps below to find the amount she saved on each item.

**1.** Subtract the Less Coupon column from the Regular Price column.

**2.** Write the difference in the Cost with Coupon column.

| *Item* | *Regular Price* | *Less Coupon* | *Cost with Coupon* |
|---|---|---|---|
| Cleanser | $1.89 | $0.39 | **$1.50** |
| Bread | $1.69 | $0.50 | ______ |
| Dog Food | $9.49 | $2.25 | ______ |
| Milk | $2.89 | $0.49 | ______ |
| Ground Beef | $2.13 | $0.25 | ______ |
| TOTALS | ______ | ______ | ______ |

*How much money did she save all together?*

**1.** Add the Regular Price column and write your answer in the blank space.

**2.** Add the Less Coupon column and write your answer in the blank space.

**3.** Subtract the Less Coupon total from the Regular Price total and write your answer in the blank space.

**Check your work by adding the Cost with Coupon column. Record your score.**

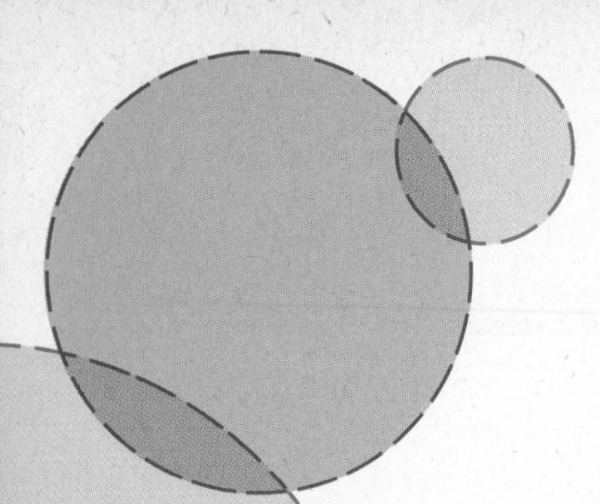

# UNIT 4

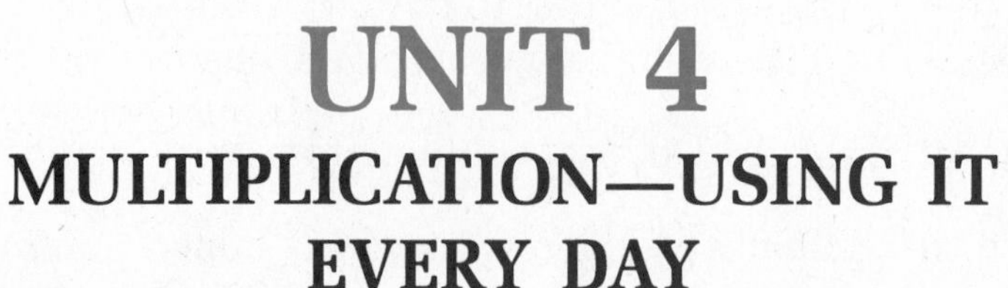

## MULTIPLICATION—USING IT EVERY DAY

### WHAT YOU WILL LEARN

When you finish this unit, you will be able to:

- Get correct answers to multiplication problems.
- Use clues to find multiplication word problems.
- Estimate answers to multiplication problems.
- Get correct answers to multiplication money problems.
- Round to get answers to multiplication problems with decimals.

## THE BASICS OF MULTIPLICATION

This is a multiplication problem:

| | | |
|---|---|---|
| 125 | Multiplicand | (also called a factor) |
| × 3 | Multiplier | (also called a factor) |
| 375 | Product | |

The **multiplicand,** the number to be multiplied (125), and the **multiplier,** the number of *times* to multiply (3), are both also called **factors.** The 125 and the 3 are factors in this problem.

The **product** is the answer to a multiplication problem (375).

### Calculator Multiplication

To solve the problem 125 × 3 = 375:

1. Enter the multiplicand, 125.
2. Press the [×] Multiplication Key.
3. Enter the multiplier, 3.
4. Press the [=] Equal Key.
5. The product, 375, should be in the answer window.

   If you did not get 375, do the problem again. Read each step slowly. Work the problem until you get 375.

6. Clear the calculator.

## Prove the Product

Factors can be entered into the calculator in any order. The product will be the same ($125 \times 3 = 375$, or $3 \times 125 = 375$).

To prove a multiplication product, multiply the multiplier by the multiplicand. For example:

| | | | | |
|---|---|---|---|---|
| 125 | Multiplicand | | 3 | Multiplier |
| × 3 | Multiplier | | × 125 | Multiplicand |
| 375 | Product | | 375 | Product |

## Horizontal Multiplication

Sometimes a multiplication problem will be written horizontally (from left to right) instead of vertically (up and down). Enter the factors the same as you would for vertical problems.

## CHECKPOINT 4-1

**YOUR GOAL:** Get 4 of 5 answers correct.

Find the products. Prove your answers. Write them in the blank spaces.

● 39 × 9 = **351** Proof: 9 × 39 = **351**

1. 42 × 7 = ______
2. 806 × 74 = ______
3. 1,643 × 835 = ______
4. 95 × 31 ______
5. 581 × 392 ______

***Check your work. Record your score.***

## SOLVING WORD PROBLEMS—MULTIPLICATION

Multiplication is repeated addition. For example, $4 \times 3$ means $4 + 4 + 4$. Four is added three times.

The clue to a multiplication problem is found by looking for a number that tells you to repeat another number a certain number of *times*.

For example, you and two friends decide to rent an apartment. Each of you must pay \$120 per month. How much is rent per month?

| Rent each Person | Person |
|---|---|
| 120 | 1 |
| 120 | 2 |
| + 120 | 3 |
| 360 | |

So 120 added together three times equals 360. A shortcut for repeating addition is to multiply a number (120) a certain number of times (3). $120 \times 3 = 360$.

## CHECKPOINT 4-2

**YOUR GOAL:** Get 4 of 5 answers correct.

Read the following problems. Find the clues that will tell you if it is a multiplication problem.

If it is a multiplication problem, write the answer in the blank space. If it is not a multiplication problem, write "No ×" in the answer space.

- Every cleaning worker at Hotel Haven cleans 25 rooms a day. Fourteen workers are on duty each day. How many rooms are cleaned each day?

  350

1. You work at Ned's Nursery. The boss tells you to find out how many oak trees are in the nursery. There are five oak trees in each bunch and there are 27 bunches. How many oak trees are there all together?

   ______

2. Jim has $17 in his wallet. Paul has $5 in his wallet. How much money do the two have together?

   ______

3. Dave is a parking lot attendant. He cannot admit more cars than can be parked in his lot. Dave counts 45 parking spaces on one row. Each row has the same number of spaces. There are nine rows. How many cars can be parked in Dave's lot?

   ______

4. Kindra's telephone bill is $34. She sends the telephone company $23. How much does Kindra owe the telephone company?

   ______

**5.** Anthony jogs four miles every day. How far will he jog in 31 days?

_______

☞ ***Check your work. Record your score.***

## CALCULATORS MAKE IT EASY!

The solar energy that comes into the solar window of your calculator powers the calculator to automatically calculate multiplication problems.

For example, if you worked the following multiplication problem on paper, you would have to do three extra steps to solve it. If you work it on your calculator, you only have to press the correct keys to get the same answer. It also takes less time to press keys than to write numbers on paper.

| **By Pencil and Paper** | | **By Calculator** | |
|---|---|---|---|
| 7,543 | Write the numbers | 7,543 | Press the numbers |
| × 691 | on paper. | × 691 | on a calculator. |
| 7,543 | You do the work | | The calculator |
| 178,870 | in your head and | | does the work |
| 4,525,800 | by hand. | | automatically |
| 5,212,213 | | 5,212,213 | |

## ROUNDING UNROUNDED PRODUCTS

Products with decimals can be unrounded. This means the answer will have as many decimal places as there are in the multiplicand plus those in the multiplier.

For example:

1. Enter the problem.

| | | Decimal Places |
|---|---|---|
| Multiplicand | 3.65 | 2 |
| Multiplier | × 2.9 | + 1 |
| Product | 10.585 | 3 |

2. The product, 10.585, is an unrounded product (10.585 has three decimal places).

An unrounded product may be more exact than you need.

To round 10.585 to two decimal places, use the method on page 38.

The rounded product is 10.59.

## CHECKPOINT 4-3

**YOUR GOAL:** Get 8 of 10 answers correct.

Find the Unrounded Product. Round each number to the number of places in the Round To column.

| | | Round To | Unrounded Product | Rounded Product |
|---|---|---|---|---|
| ● | 3.28 × 57.6 | 2 places | **188.928** | **188.93** |
| **1.** | 7.43 × 8.02 | 2 places | ______ | ______ |
| **2.** | 9.138 × 0.647 | 3 places | ______ | ______ |
| **3.** | 0.6051 × 85.27 | 3 places | ______ | ______ |
| **4.** | 74.326 × 0.34 | 4 places | ______ | ______ |
| **5.** | 2.8703 × 9.425 | 4 places | ______ | ______ |

☞ ***Check your work. Record your score.***

## AMOUNTS OF MONEY IN MULTIPLICATION

Calculators make multiplying amounts of money easy because the calculator places the decimals for you. Just press the [·] Decimal Point Key in all factors where there is a decimal. Enter **6.7 × 3.21** = into your calculator. The product, 21.507, should be in the answer window.

Many business and personal math problems are about money (dollars and cents) and have only two decimal places. When you multiply amounts of money, you need only two decimal places in the product. If the product in the answer window has more than two decimal places, round to get a number that has two decimal places.

To round $9.43 × 7.6 = $71.668:

1. Count three places to the right of the decimal (8).
2. If the digit is
   a. *5 or more,* add 1 to the number to the left of that digit (8 is *5 or more,* so add 1 + 6 = 7).
   b. *less than 5,* keep the same digit in the place to the left.
3. The rounded product is $71.67.

## CHECKPOINT 4-4

**YOUR GOAL:** Get 4 of 5 answers correct.

Find the product. Round it to two decimal places.

● Jill fills her car with 12.5 gallons of gasoline. The price of gasoline is $1.309. How much will Jill pay for the gasoline?

**$16.17**

**1.** At the supermarket you buy 4.2 pounds of meat. The price of meat is $3.17 a pound. How much will you pay?

______

**2.** Troy's bedroom needs painting. Paint costs $15.95 a gallon. He needs four gallons of paint. How much will Troy pay for the paint?

______

**3.** Jack makes $4.25 an hour. He worked 36.5 hours. How much will Jack be paid?

______

**4.** Roger wants to buy new tires for his car. One new tire costs $89.95. Roger buys four tires. How much will he pay?

______

**5.** Dana Brown buys 12 calves from a livestock sale. Each calf costs $450. How much will farmer Brown pay?

______

***Check your work. Record your score.***

## ESTIMATING PRODUCTS

Estimating multiplication products helps you know if the product seems correct.

For example, you estimate that a product should be near 600, but you get 6,390 for an answer. You know that you either estimated wrong or calculated wrong. Rework the problem to get the correct answer.

To estimate the product for $64 \times 9$:

1. Round the factors to the highest place at the left (64 becomes 60, and 9 becomes 10). Write the rounded factors in the Rounded Factors column.
2. Multiply the rounded factors on your calculator. Write the product (600) in the Estimated Product column.
3. Use your calculator to multiply the original factors (64 × 9). Write the total (576) in the Calculator Product column.
4. If the Calculator Product is not close to the Estimated Product, estimate and calculate the answer again. Write the correct Estimated Product and the Calculator Product.

## *CHECKPOINT 4-5*

**YOUR GOAL:** Get 12 of 15 answers correct.

Round the factors. Find the Estimated Products and the Calculator Products.

| | Original Factors | Rounded Factors | Estimated Product | Calculator Product |
|---|---|---|---|---|
| ● | 64 × 9 | 60 × 10 | 600 | 576 |
| **1.** | 86 × 32 | | | |
| **2.** | 395 × 61 | | | |
| **3.** | 457 × 308 | | | |
| **4.** | 4,067 × 48 | | | |
| **5.** | 8,962 × 409 | | | |

 ***Check your work. Record your score.***

### JUST FOR FUN

Write the number answer in the Answer column. Turn your calculator upside down and write the word in the Word column.

| | | Answer | Word |
|---|---|---|---|
| **1.** | 3,869 × 2 = | | |
| **2.** | 169 × 2 = | | |
| **3.** | 551 × 14 = | | |
| **4.** | 151 × 4 = | | |

## WHAT YOU HAVE LEARNED

Now that you have completed this unit, you should be able to:

- Multiply whole numbers correctly.
- Use clues to find multiplication word problems.
- Estimate multiplication answers.
- Multiply amounts of money correctly.
- Round decimal answers in multiplication problems.

# PUTTING IT TOGETHER

## ACTIVITY 4-1 YOUR GOAL: Get 4 of 5 answers correct.

Find the products.

● 73 × 51 = **3,723**

1. 82 × 65 = ______
2. 821 × 57 = ______
3. 1,840 × 5,297 = ______
4. 976 × 51 = ______
5. 6,439 × 704 = ______

***Check your work. Record your score.***

## ACTIVITY 4-2 YOUR GOAL: Get 4 of 5 answers correct.

Find the products.

● John drives 18 miles to work and back home every day. He traveled to work 23 days in April. How many miles did John travel to work and back home in April?

**414**

1. Henry feeds his cattle seven bags of grain a week. Each bag weighs 50 pounds. How many pounds of grain does Henry feed a week?

______

2. Paula employs 12 people. Each person works 40 hours a week. How many hours do Paula's employees work each week?

______

3. Randall Recycling volunteers collect newspaper in bundles. They put 15 pounds of paper in each bundle. The volunteers collected 124 bundles during one drive. How many pounds of paper did the volunteers collect?

______

4. Mike used 25 boxes of screws while working on his construction job. Each box has 150 screws. How many screws did Mike use?

______

5. The community theater sold out all six nights of its performances. The theater seats 525 people. How many people saw the play?

________

☞ *Check your work. Record your score.*

## ACTIVITY 4-3 YOUR GOAL: Get 8 of 10 answers correct.

Find the Unrounded Product. Round it to the number of places in the Round To column.

| | | Round To | Unrounded Product | Rounded Product |
|---|---|---|---|---|
| ● | 48.9 × 0.33 | 2 places | 16.137 | 16.14 |
| 1. | 2.04 × 0.052 | 2 places | | |
| 2. | 0.136 × 0.457 | 3 places | | |
| 3. | 7.1916 × 44.5 | 3 places | | |
| 4. | 30.973 × 4.02 | 4 places | | |
| 5. | 70.3084 × 4.2 | 4 places | | |

☞ *Check your work. Record your score.*

## ACTIVITY 4-4 YOUR GOAL: Get 4 of 5 answers correct.

Find the product. Round it to two decimal places.

● Gasoline is $1.329 per gallon. You put 13.5 gallons of gasoline in your car. How much will you owe?

**$17.94**

1. Keisha bought 16 loaves of bread for a fund-raising supper. Each loaf costs $1.36. How much did Keisha spend on bread?

________

2. Pat wants to put tile on her kitchen floor. The tiles cost $7.65 a square yard. Pat needs nine square yards. How much will Pat pay?

________

3. Carlos earns $5.25 per hour. He worked 37.25 hours. How much did he earn?

______

4. Darin needs to buy home heating oil. Oil costs $1.32 per gallon. He buys 252.5 gallons. How much does Darin owe?

______

5. Mohan Movie Theater seats 175 people. On Friday night all of the seats were filled. Each seat costs $5. How much did the movie theater collect?

______

☞ ***Check your work. Record your score.***

## ACTIVITY 4-5 **YOUR GOAL:** Get 12 of 15 answers correct.

Give the Rounded Factors, the Estimated Products, and the Calculator Products.

| | | Rounded Factors | Estimated Product | Calculator Product |
|---|---|---|---|---|
| ● | 88 × 23 | 90 × 20 | 1,800 | 2,024 |
| 1. | 74 × 96 | ______ | ______ | ______ |
| 2. | 510 × 87 | ______ | ______ | ______ |
| 3. | 375 × 224 | ______ | ______ | ______ |
| 4. | 1,852 × 119 | ______ | ______ | ______ |
| 5. | 7,485 × 4,261 | ______ | ______ | ______ |

☞ ***Check your work. Record your score.***

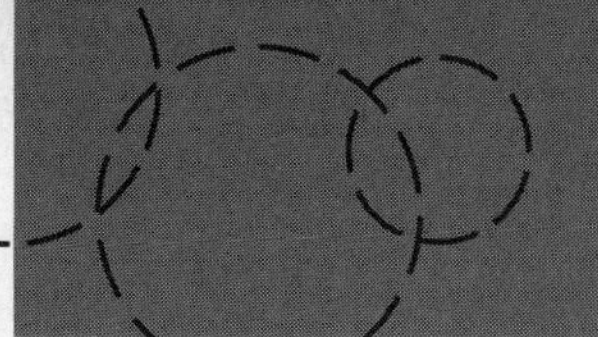

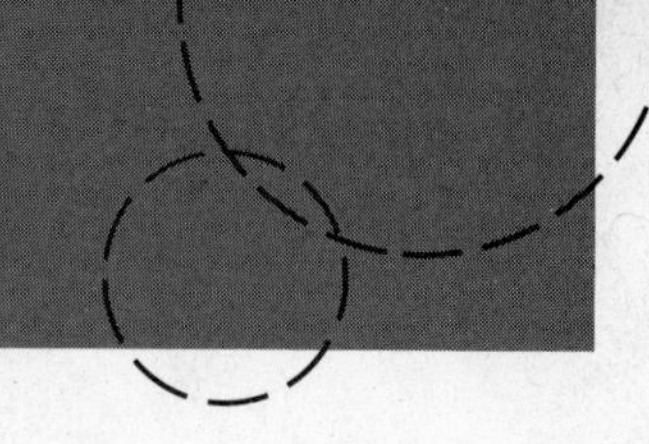

# MAKING IT WORK

## REVIEW 4-1 YOUR GOAL: Get 5 of 6 answers correct.

### Inventory

When a person wants to know how many items he or she has to sell or use, an **inventory** of items on hand is taken. An inventory is a complete list of the number and type of each item in stock. For example:

Fred sells fresh fruit and vegetables. He decided to take inventory of his produce. He filled in the following chart.

Find out how many of each fruit and vegetable Fred has. Multiply the number of baskets of fruit (5 baskets of apples) by the number of pieces of fruit in each basket (12 apples per basket).

Fred's Fresh Produce — Date: August 7, 1992

| Vegetable/Fruit | Number per Basket | Number of Baskets | Total |
|---|---|---|---|
| Apples | 12 | 5 | 60 |
| Tomatoes | 12 | 15 | ______ |
| Cucumbers | 4 | 11 | ______ |
| Carrots | 12 | 9 | ______ |
| Cantelope | 2 | 19 | ______ |
| Corn | 6 | 21 | ______ |
| Onions | 12 | 17 | ______ |

☞ ***Check your work. Record your score.***

# UNIT 5
## DIVISION—USING IT EVERY DAY

### WHAT YOU WILL LEARN

When you finish this unit, you will be able to:

- Get correct answers to division problems.
- Use clues to find division word problems.
- Round to get answers to division problems with decimals.
- Estimate answers to division problems.
- Get correct answers to division money problems.
- Get correct answers to chain division problems.

## THE BASICS OF DIVISION

This is a division problem:

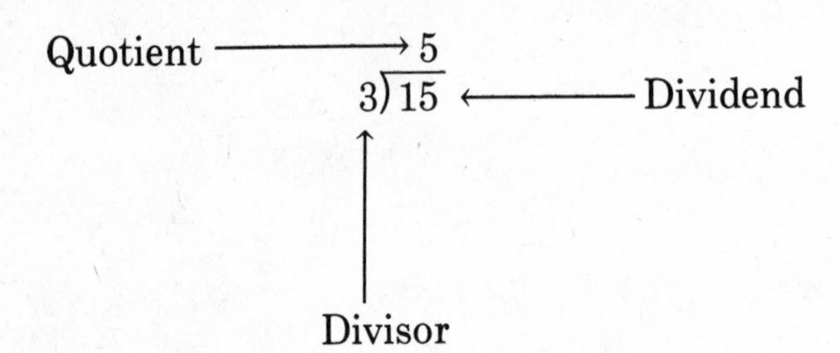


The **dividend** (15) is the number to be divided. The **divisor** (3) is the number that does the dividing. The **quotient** (5) is the answer to a division problem.

### Writing Division Problems

There are three ways to write a division problem:

| | | |
|---|---|---|
| $\text{Divisor}\overline{)\text{Dividend}}$ with Quotient above<br>$3\overline{)15}$ with 5 above | Dividend ÷ Divisor = Quotient<br>$15 \div 3 = 5$ | $\frac{\text{Dividend}}{\text{Divisor}} = \text{Quotient}$<br>$\frac{15}{3} = 5$ |

In each case, always enter the dividend first. Then enter the divisor into your calculator.

## Calculator Division

To solve the problem 92 ÷ 4 = 23:

1. Enter the dividend, 92.
2. Press the [÷] Division Key. In your mind, say "divided by."
3. Enter the divisor, 4.
4. Press the [=] Equal Key.
5. The quotient, 23, should be in the answer window.

   If you did not get 23, work the problem until you get 23. Read each step slowly.
6. Clear the calculator.

## Prove the Quotient

To prove a division quotient, multiply the quotient by the divisor. For example:

23 × 4 = 92

If the answer is the same as the dividend, the quotient is correct. If the answer is different, work the problem again until the proof answer is the same as the dividend.

## CHECKPOINT 5-1

**YOUR GOAL:** Get 4 of 5 answers correct.

Find the quotients. Prove your answers. Write them in the blank spaces.

● 108 ÷ 12 = 9 Proof: 9 × 12 = 108

1. 338 ÷ 13 = ______
2. 2,730 ÷ 15 = ______
3. 10,404 ÷ 68 = ______
4. 8,432 ÷ 34 = ______
5. 12,194 ÷ 469 = ______

***Check your work. Record your score.***

## SOLVING DIVISION WORD PROBLEMS

Division is finding out how many equal parts are in a whole. Division takes a whole item, such as a piece of cloth, and "cuts" it into equal parts.

The clue to a division problem is found by looking for a whole that needs to be cut into equal parts.

You picked 36 apples from a tree in your yard. You decide to bring them to class and divide them between your classmates. There are 12 students in your class. You want to give each classmate the same number of apples. How many apples will each classmate get?

This is a division problem because a number (the dividend) is to be divided by another number (the divisor).

The trick to solving a division word problem is figuring out which is the dividend and which is the divisor. After you find the dividend and the divisor, you can complete the division problem and find the quotient. For example, in the apple problem the dividend is the item you are going to divide up: the apples! The divisor is the number of people who will get the apples. The quotient is the number of apples each classmate will get. From THE BASICS OF DIVISION, you know that:

Dividend ÷ Divisor = Quotient

so:

36 apples ÷ 12 classmates = *3 apples per person*

### CHECKPOINT 5-2

**YOUR GOAL:** Get 4 of 5 answers correct.

Read the following problems. Find the clues that tell you if it is a division problem.

If it is a division problem, write the answer in the blank space. If it is not a division problem, write "No ÷ " in the answer space.

● Will's Window Cleaning Service has to clean the windows on the local bank. There are 960 windows on the bank building. Will has 15 employees. How many windows will each employee have to clean?

**64**

**1.** Jimmy has to plow a 30-acre field. He decides he can plow five acres an hour. How many hours will it take for Jimmy to plow the field?

________

**2.** Penny buys six cans of juice for $1.06 each. How much will Penny pay for the juice?

________

**3.** Kelvin drove 274 miles. He then filled up the tank with 13.7 gallons of gasoline. How many miles a gallon does Kelvin's car get?

________

**4.** There are three boxes of fruit in the storeroom. One box has five cans. One box has seven cans. The other box has 13 cans. How many cans of fruit are in the storeroom?

________

**5.** A road crew needs to fix 2,000 feet of road. They want to put the caution barrels five feet apart. How many caution barrels does the road crew need?

________

☞ ***Check your work. Record your score.***

## CALCULATORS MAKE IT EASY!

Calculators make division easier than doing the problem by pencil and paper. On a calculator you do not have to do as many steps to get an answer. For example:

To work the problem 1,965 ÷ 5, you have to do these steps to get the answer.

By Pencil and Paper

```
   393
5)1965
  15
   46
   45
    15
    15
     0
```

Write the numbers on paper. You do the work in your head and by hand.

By Calculator

```
   393
5)1965
```

Press the numbers, and the calculator does the work automatically.

## ROUNDING UNROUNDED QUOTIENTS

Quotients with decimals can be unrounded. This means the answer will have as many decimal places as appear in the answer window.

For example:

1. Enter the problem.

   65 ÷ 8 = 8.125

2. The quotient, 8.125, is an unrounded quotient (8.125 has three decimal places).

An unrounded quotient may be more exact than you need.

To round 8.125 to the nearest whole number, follow the instructions in Unit 2.

The rounded quotient is 8.13.

## CHECKPOINT 5-3

**YOUR GOAL:** Get 8 of 10 answers correct.

Find the Unrounded Quotient. Round each number to the number of places in the Round To column.

| | Round To | Unrounded Quotient | Rounded Quotient |
|---|---|---|---|
| ● 6.58 ÷ 4 | 2 places | 1.645 | 1.65 |
| **1.** 79 ÷ 8 | 2 places | ________ | ________ |
| **2.** 96.3 ÷ 8 | 3 places | ________ | ________ |
| **3.** 339 ÷ 7 | 3 places | ________ | ________ |
| **4.** 257 ÷ 14 | 4 places | ________ | ________ |
| **5.** 452 ÷ 102 | 4 places | ________ | ________ |

☞ ***Check your work. Record your score.***

## ESTIMATING QUOTIENTS

Estimating division quotients helps you know if the answer seems correct.

For example, if you estimate that a quotient should be near 600 but you get 62 for an answer, you know you either estimated wrong or calculated wrong. Rework the problem to get the correct answer.

To estimate the quotient for 294 ÷ 21:

1. Round the numbers to the highest place at the left (294 becomes 300, and 21 becomes 20). Write the rounded factors in the Rounded Factors column.
2. Divide the rounded numbers on your calculator. Write the quotient (15) in the Estimated Quotient column. Round to the nearest whole number.
3. Use your calculator to divide the original numbers (294 ÷ 21). Write the quotient (14) in the Calculator Product column.
4. If the Calculator Product is not close to the Estimated Product, estimate and calculate the answer again. Write the correct Estimated Product and the Calculator Quotient.

## CHECKPOINT 5-4

**YOUR GOAL:** Get 12 of 15 answers correct.

Round the numbers. Find the Estimated Quotients and the Calculator Quotients.

| | Original Numbers | Rounded Numbers | Estimated Quotient | Calculator Quotient |
|---|---|---|---|---|
| ● | 294 ÷ 21 | 300 ÷ 20 | 15 | 14 |
| 1. | 192 ÷ 24 | | | |
| 2. | 1,768 ÷ 52 | | | |
| 3. | 2,856 ÷ 34 | | | |
| 4. | 4,564 ÷ 326 | | | |
| 5. | 32,660 ÷ 1,420 | | | |

☞ ***Check your work. Record your score.***

## AMOUNTS OF MONEY IN DIVISION

Calculators automatically place a decimal at the correct place in a quotient. This makes dividing amounts of money easy.

Many business and personal math problems are about money (dollars and cents) that has only two decimal places. When you divide amounts of money, you need only two decimal places in the quotient. If the quotient in the answer window has more than two decimal places, round to two decimal places.

To round \$1.98 ÷ 32 = 0.061875, follow the instructions in Unit 2.

The rounded quotient is \$0.06.

## CHECKPOINT 5-5

**YOUR GOAL:** Get 4 of 5 answers correct.

Find the quotient. Round it to two decimal places.

● Apples cost \$.70 a pound. How many pounds can you buy with \$3.50?

**5**

1. Shampoo costs \$.98. There are 22 ounces in one bottle of shampoo. How much does the shampoo cost an ounce?

________

**2.** Martin brought 2,500 bushels of wheat to the market. He received $5,250 for the wheat. How much money did Martin get per bushel for his wheat?

________

**3.** Four friends rent an apartment. The rent is $525 a month. How much will each friend have to pay per month?

________

**4.** Gasoline costs $1.349 a gallon. Karmen's car gets 22 miles to the gallon. How much does gasoline cost per mile for Karmen?

________

**5.** Charles bought eight spark plugs. He wrote the cost of $15.84 in his checkbook. Jim asked Charles to sell him a spark plug. How much should Charles charge Jim for one spark plug?

________

***Check your work. Record your score.***

## CHAIN DIVISION

Your calculator makes it possible to divide a quotient by another divisor one or more times. This is called **chain division.**

To work: 126 ÷ 24 ÷ 6 = 0.88

1. Enter the dividend (126), and press [÷].
2. Enter the divisor (24), and press [÷].
3. Enter the second divisor (6), and press [=].
4. Round the answer (0.875) to two decimal places.
5. Write the rounded quotient (0.88) in the blank space.

## CHECKPOINT 5-6

**YOUR GOAL:** Get 4 of 5 answers correct.

Use chain division to solve the problems. Round the answer to two decimal places.

● 728 ÷ 7 ÷ 8 = 13

**1.** 705 ÷ 20 ÷ 25 = ________

**2.** 1,344 ÷ 48 ÷ 15 = ________

**3.** 13,167 ÷ 627 ÷ 3 = ________

**4.** 386,274 ÷ 119 ÷ 14 = _______

**5.** 554,268 ÷ 34 ÷ 627 = _______

☞ ***Check your work. Record your score.***

## JUST FOR FUN

Work the problem and turn the calculator upside down to read the answer to the question.

1. What does a person do when he or she has a broken leg?

   1,515,216 ÷ 4 = _______
2. Sometimes your blood does this.

   42,648 ÷ 6 = _______
3. If a Thanksgiving dinner could talk, what would it say?

   757,612 ÷ 2 = _______
4. This is what you wear over your sock.

   45,675 ÷ 15 = _______
5. You do not want to fall into one of these.

   29,632 ÷ 8 = _______

## WHAT YOU HAVE LEARNED

Now that you have completed this unit, you should be able to:

- Divide whole numbers correctly.
- Use clues to find division word problems.
- Round decimal answers in division problems.
- Estimate division answers.
- Divide amounts of money correctly.
- Solve chain division problems.

# PUTTING IT TOGETHER

## ACTIVITY 5-1 YOUR GOAL: Get 4 of 5 answers correct.

Find the quotients. Do not round.

- ● 42 ÷ 8 = 5.25
1. 85 ÷ 4 = ______
2. 307 ÷ 8 = ______
3. 568 ÷ 16 = ______
4. 511 ÷ 73 = ______
5. 26,214 ÷ 771 = ______

☞ ***Check your work. Record your score.***

## ACTIVITY 5-2 YOUR GOAL: Get 4 of 5 answers correct.

Find the quotients.

- ● Emitt feeds his cattle 35,000 pounds of grain a week. He feeds the same amount of grain each day. How many pounds of grain does Emitt feed his cattle each day?

  5,000

1. Christine is making chicken nuggets at Casy's Restaurant. A serving is six nuggets. She cooks a case of nuggets. One case has 480 nuggets in it. How many servings are prepared for sale?

   ______

2. Marcy puts cheese in her sandwiches for work each day. She buys a package of cheese that has 75 slices in it. Marcy uses 1.5 pieces of cheese a day. How many days will Marcy have cheese for her sandwiches?

   ______

3. The Cleaning Service has 525 rags for their workers to use. There are 25 workers on the Cleaning Service staff. How many rags will each worker get to use?

   ______

4. Randy made six trips to work and back home. He has gone 102 miles. How far is Randy's trip to work and back home?

________

5. Toby needs 675 screws to finish a shop job. Each box has 75 screws in it. How many boxes does Toby need to finish the job?

________

☞ ***Check your work. Record your score.***

## ACTIVITY 5-3 YOUR GOAL: Get 12 of 15 answers correct.

Give the Estimated Quotients and the Calculator Products.

| | Original Numbers | Rounded Numbers | Estimated Quotient | Calculator Quotient |
|---|---|---|---|---|
| ● | 95 ÷ 5 | 100 ÷ 10 | 10 | 19 |
| 1. | 235 ÷ 5 | | | |
| 2. | 611 ÷ 47 | | | |
| 3. | 5,706 ÷ 9 | | | |
| 4. | 3,345 ÷ 15 | | | |
| 5. | 49,956 ÷ 724 | | | |

☞ ***Check your work. Record your score.***

## ACTIVITY 5-4 YOUR GOAL: Get 4 of 5 answers correct.

Find the quotient. Round it to two decimal places.

● George works at a carpentry shop. He and two other workers built a cabinet. They sold the cabinet and had a profit of $96. How much money will each of the three get for the cabinet?

**$32**

1. Kami bought seven square yards of carpet for a room in her home. Her husband wants to know how much she paid per square yard for the carpet. She knows she paid $68.95 for all of the carpet. How much did she pay per square yard?

________

**2.** Crystallia worked 37 hours last week. She got paid $175.75. How much does Crystallia earn per hour?

________

**3.** Amanda wants to buy her four children small rewards for getting good grades in school. She only has $10 to spend. How much can she spend on each child?

________

**4.** Kirk delivered 36,500 pounds of tomatoes to the produce terminal. If the tomatoes were packed in buckets of 50 pounds each, how many buckets of tomatoes were delivered?

________

**5.** Melissa puts a microwave oven on layaway. After she pays the down payment, she owes $276.82. She plans to pay it off in eight months. How much will she have to pay each month?

☞ ***Check your work. Record your score.***

## ACTIVITY 5-5 **YOUR GOAL:** Get 8 of 10 answers correct.

Find the Unrounded Quotient. Round each number to the number of places in the Round To column.

| | | Round To | Unrounded Quotient | Rounded Quotient |
|---|---|---|---|---|
| ● | 34 ÷ 16 | 2 places | 2.125 | 2.13 |
| **1.** | 108.9 ÷ 12 | 2 places | ________ | ________ |
| **2.** | 572 ÷ 6 | 3 places | ________ | ________ |
| **3.** | 6.75 ÷ 4 | 3 places | ________ | ________ |
| **4.** | 484 ÷ 73 | 4 places | ________ | ________ |
| **5.** | 896 ÷ 405 | 4 places | ________ | ________ |

☞ ***Check your work. Record your score.***

## ACTIVITY 5-6 YOUR GOAL: Get 4 of 5 answers correct.

Use chain division to solve the problems.

- ● 291 ÷ 30 ÷ 2 = 4.85
1. 810 ÷ 18 ÷ 5 = ______
2. 276 ÷ 24 ÷ 23 = ______
3. 6,844 ÷ 58 ÷ 2 = ______
4. 5,375 ÷ 43 ÷ 25 = ______
5. 27,648 ÷ 36 ÷ 3 = ______

☞ ***Check your work. Record your score.***

# PART TWO
## COMBINED AND MEMORY OPERATIONS

## UNIT 6
### COMBINED OPERATIONS—USING THEM EVERY DAY

## UNIT 7
### MEMORY OPERATIONS—USING THEM EVERY DAY

# UNIT 6

# COMBINED OPERATIONS—USING THEM EVERY DAY

## WHAT YOU WILL LEARN

When you finish this unit, you will be able to:

- Get correct answers to combined operation problems.
- Solve combined operation word problems.
- Find averages.
- Estimate answers to combined operation problems.

## THE BASICS OF COMBINED OPERATIONS

**Combined operations** are two or more math operations (addition, subtraction, multiplication, or division) used to solve a problem.

Operations can be done in sequence, which means you do not have to clear the calculator between parts of the problem.

For example, work the problem $\mathbf{4 \times 3 \div 2 - 15 + 17}$ = ________ on your calculator. Do not clear the calculator between operations. Your answer should be 8.

## *CHECKPOINT 6-1*

YOUR GOAL: Get 4 of 5 answers correct.

Use combined operations to solve the problems. Write your answers in the blank spaces. Do not round.

| | | | | | |
|---|---|---|---|---|---|
| ● | $3 \times 82$ | $+ 51$ | | = | 297 |
| **1.** | $85 + 69$ | $+ 89$ | $+ 67 \div 4$ | = | ________ |
| **2.** | $3 \times 35$ | $\times 4$ | $\div 16$ | = | ________ |
| **3.** | $85 \times 3$ | $+ 76$ | $\div 4$ | = | ________ |

**4.** 247 − 35 − 85 − 28 + 79 = ________

**5.** 50 × 1.09 × .80 = ________

☞ ***Check your work. Record your score.***

## YESTERDAY, TODAY, AND TOMORROW

Some restaurants are using hand-held devices to take orders from customers. Each item on the menu has a special code that tells the computer what has been ordered. For example, hotcakes would be coded HOTC. After the order is put into the hand-held device, the waiter presses a key and the order is sent to a computer. The cook then presses a button and the order is printed on a piece of paper, which is followed to prepare the meal. When customers finish their meals, the waiter gives them printed copies of their orders, which are used to pay the cashier.

## SOLVING WORD PROBLEMS—COMBINED OPERATIONS

Some word problems can be solved using combined operations. To work a combined operation word problem, you must first find out what the problem is asking you to solve. Then decide how to set up the problem.

For example:

Shelly has a checking account. Her beginning June balance was $257.76. In June, Shelly **withdrew** (subtracted) $35.65, $86.50, and $41.35. She also **deposited** (added) $264.56. What was Shelly's ending balance for June?

Follow these steps to solve the word problem.

1. Record key words and numbers.

| | |
|---|---|
| June balance: | $257.76 |
| Withdrawals: | 35.65 |
| | 86.50 |
| | 41.35 |
| Deposit: | 264.56 |

2. Find Shelly's ending balance.

   | Beginning Balance | Withdrawals | | | Deposit |
   |---|---|---|---|---|
   | 257.76 | − 35.65 | − 86.50 | − 41.35 | + 264.56 |

3. Record your answer ($358.82).

To solve a combined operation word problem you may need to use multiplication or division as well as addition and subtraction. For example:

The mechanics at Jim's Service Station use 15 bottles of Premium Oil a week. Jim wants to order enough oil for a month (four weeks). Each box of Premium Oil contains 12 bottles. How many boxes does Jim need to order?

1. Record key words and numbers.

   15 bottles per week
   4 weeks
   12 bottles per box

2. Find how many boxes Jim needs to order.

   15 (bottles per week)
   × 4 (weeks)
   60 (bottles for 4 weeks) ÷ 12 (bottles per box) = 5 (boxes)

3. Record your answer.

## CHECKPOINT 6-2

**YOUR GOAL:** Get 4 of 5 answers correct.

If necessary, write the key words and numbers on a piece of paper. Then work the problems on your calculator. Write your answers in the blank spaces.

● Ray's beginning March balance is $598.64. He withdrew $75.40, $20.65, $35.24, and $231.75. He also deposited $40.78, and $367.25. What is Ray's ending March balance?

$598.64 − $75.40 − $20.65 − $35.24 − $231.75 + $40.78 + 367.25 =

**$643.63**

**1.** James buys five bales of hay. The bales cost $350 per ton. The bales weigh: 0.9 ton, 1 ton, 1.1 ton, 0.8 ton, 1 ton. How much will James pay for all of the bales? (*Hint:* First add the weight of each bale to get the total tons. Next multiply the total tons by the cost per ton.)

________

**2.** You go to the grocery store and buy these items:

| | |
|---|---|
| toothpaste | $1.33 |
| milk | $2.15 |
| shampoo | $1.45 |

You hand the cashier a $20 bill. How much change should you get back?

______

**3.** Stephen gathers aluminum cans from his neighborhood to take to the Waste Saver Recycling Center. He collected the following amounts of cans during the week:

| | |
|---|---|
| Monday | 5 pounds |
| Wednesday | 3.5 pounds |
| Friday | 6.5 pounds |

The center pays $0.35 a pound. How much will Stephen be paid for the cans?

______

**4.** Each cleaning worker at First Bank uses two cans of glass cleaner a week. There are 14 cleaning workers who work at First Bank. Tico has to order enough glass cleaner for four weeks. Each box of glass cleaner holds 16 bottles. How many boxes of glass cleaner will Tico have to order?

______

**5.** Last week Sarah worked:

| | |
|---|---|
| Monday | 5.5 hours |
| Tuesday | 7 hours |
| Wednesday | 6 hours |
| Thursday | 7 hours |
| Friday | 6.5 hours |

Sarah earns $4.25 an hour. How much did Sarah earn last week?

______

☞ ***Check your work. Record your score.***

## AVERAGES

To find an **average**:

1. *Add* the numbers you want to average to get the total.
2. *Divide* the total by the number of addends.
3. *Record* your answer.

For example:

| Numbers to be Averaged | Total | | Number of Addends | | Average |
|---|---|---|---|---|---|
| 24 + 36 + 33 = | 93 | ÷ | 3 | = | 31 |

Many average problems will be word problems. For example:

Jan has taken three tests in her math class. Her grades are 75, 85, 83. What is her math average?

1. Add all of the grades together.

   75 + 85 + 83 = 243
2. Divide that total (243) by the number of grades (3).

   243 ÷ 3 = 81.
3. Jan's math average is 81.

## CHECKPOINT 6-3

**YOUR GOAL:** Get 4 of 5 answers correct.

Find the averages. Write your answers in the blank spaces.

● Maria has taken six tests in her English class. Her grades are: 95, 89, 82, 64, 83, 81. What is Maria's English average, rounded to the nearest whole number?

82

**1.** Betty works at Kathy's Klothes. On Monday she sold six dresses. Tuesday she sold nine dresses. Wednesday she sold 15 dresses. What is the average number of dresses Betty sold each day?

________

**2.** Kaye is making out her food budget for May. She will use an average of the cost of food from the last four months to make her May budget. Her food costs for the past four months were:

| | |
|---|---|
| January | $126.75 |
| February | $164.29 |
| March | $172.95 |
| April | $159.45 |

What is Kaye's average food cost for the past four months?

________

**3.** You put 15.7 gallons of gasoline in your car and traveled 315 miles. The second time you put 14.3 gallons of gasoline in your car and traveled 285 miles. What is the average number of miles your car will go on one gallon of gasoline? (*Hint*: 1. Add the number of gallons of gasoline and get a total. 2. Add the number of miles and get a total. 3. Divide the total miles by the total gallons of gasoline and get the average number of miles.)

________

**4.** Rainfall has been below average this year.

| | |
|---|---|
| January | 1.34 inches |
| February | 1.68 inches |
| March | 1.05 inches |
| April | 2.98 inches |
| May | 3.50 inches |
| June | 2.41 inches |

What is the average monthly rainfall for the first six months of the year?

________

**5.** In baseball, to find a batting average divide the number of hits by the number of times at bat. Jackson played in three games and was at bat 16 times. He made three hits in the first game, two hits in the second game, and two hits in the third game. What is Jackson's batting average?

________

☞ ***Check your work. Record your score.***

## JUST FOR FUN

1. What letter is in the word igloo and at the end of leg? ________

   240 ÷ 8 ÷ 5 = ________

2. What does Santa Claus arrive in? ________

   949,906 ÷ 2 − 13,578 = ________

3. What do people do when something is funny? ________

   567,241 + 753,232 − 376,616 − 567,241 = ________

Check your work.

## ESTIMATING—COMBINED OPERATIONS

Estimating helps you to decide if your answer seems correct. For example:

To estimate 27 × 11 + 36 = ______

1. Round each number to the highest place at the left (27 becomes 30, 11 becomes 10, and 36 becomes 40). Write the rounded problem in the Rounded Problem column.
2. Work the rounded problem mentally. Write the rounded answer (340) in the Estimated Answer column.
3. Use your calculator to work the original problem (27 × 11 + 36). Write the total in the Calculator Answer column (333).
4. If the Calculator Answer is not close to the Estimated Answer, estimate and calculate the answer again. Write the correct Estimated Answer and the Calculator Answer.

| Original Problem | Rounded Problem | Estimated Answer | Calculator Answer |
|---|---|---|---|
| 27 × 11 + 36 | 30 × 10 + 40 | 340 | 333 |

### CHECKPOINT 6-4

**YOUR GOAL:** Get 12 of 15 answers correct.

Find the rounded problems, use them to get the Estimated Answers. Then find the Calculator Answers. Write your answers in the blank spaces.

| | Original Problem | Rounded Problem | Estimated Answer | Calculator Answer |
|---|---|---|---|---|
| ● | 27 × 11 + 36 = | 30 × 10 + 40 | 340 | 333 |
| **1.** | 215 × 3 + 17 = | ______ | ______ | ______ |
| **2.** | 47.98 × .2 + 47.98 = | ______ | ______ | ______ |
| **3.** | 22 + 10 ÷ 2.13 = | ______ | ______ | ______ |
| **4.** | 45 + 78 + 84 ÷ 3 = | ______ | ______ | ______ |
| **5.** | 2,032 − 89 − 214 + 23 = | ______ | ______ | ______ |

☞ ***Check your work. Record your score.***

## WHAT YOU HAVE LEARNED

Now that you have completed this unit, you should be able to:

- Get correct answers to combined operation problems.
- Set up and solve combined operation word problems.
- Find averages.
- Estimate answers to combined operations problems.

# PUTTING IT TOGETHER

## ACTIVITY 6-1 YOUR GOAL: Get 4 of 5 answers correct.

Use combined operations to solve the problems. Write your answers in the blank spaces.

- ● 256 − 32 ÷ 16 = **14**

1. 1.45 + 2.15 + 1.75 + 4.25 ÷ 4 = ______
2. 794 + 75 − 34 − 97 − 184 = ______
3. .50 × 775 × .70 = ______
4. 259 − 10 + 25 = ______
5. 6 × 120 ÷ 12 = ______

☞ ***Check your work. Record your score.***

## ACTIVITY 6-2 YOUR GOAL: Get 4 of 5 answers correct.

If necessary, write the key words and numbers on a piece of paper. Then work the problems on your calculator. Write your answers in the blank spaces.

- ● Melissa wants to take a bus to see her sick grandmother in Amarillo. The price for the bus ticket was $16 before a new tax that raised the price $.05 per dollar. What is the price for the ticket now?

  $16 × $.05 + $16 = **$16.80**

1. Christie earns $152 a week. There are four weeks in a month. Rent is $285 a month. How much money will Christie have to spend each month on other bills after she pays her rent?

   ______

2. Willie's Window Cleaning Service employs 12 people. Willie recently won a contract to clean the windows on a six-floor building. Each floor of the building has 120 windows. Willie wants to give the same number of windows to each worker. How many windows will each worker have to clean?

   ______

3. Pam's beginning checking account balance for November was $378.42. In November she withdrew $75.50, $24.86, $10.45, and $137.84. She also deposited $347.65. What was Pam's ending November balance?

   ______

4. Lawrence was told by his boss that the warehouse needed 3,000 light bulbs. There were only 2,000 in stock. An order of 600 light bulbs that cost $300 came in. How much will it cost for 400 more bulbs? (*Hint*: Divide the $300 cost by the 600 light bulbs. Then multiply the quotient by 400.)

________

5. Mr. Kimbro is planning to irrigate his four fields. Each field has 120 rows. He will irrigate 30 rows a day. How many days will it take to irrigate Mr. Kimbro's farm?

________

☞ ***Check your work. Record your score.***

## ACTIVITY 6-3 YOUR GOAL: Get 4 of 5 answers correct.

Find the averages. Write your answers in the blank spaces.

● The weekly salaries for the cashiers at The Building Supply Shop are: $220, $260, $154, and $190. What is the average salary paid to each cashier?

**206**

1. Stephanie went bowling and scored: 176, 124, and 171. What was Stephanie's bowling average?

________

2. Chris is making out a budget for April. His electricity bills for the past three months were: $35.76, $42.55, and $24.62. What is Chris's average monthly cost for electricity?

________

3. Cindy is starting a new diet. Before she begins, she must figure the average amount of calories she takes in each day. She writes down these totals for a seven-day period: 2,146; 1,976; 1,818; 1,738; 1,957; 1,589; and 2,042. What was the average amount of calories Cindy took in each day?

________

4. A flood hit the Pedro's farm. It rained 3.25 inches the first day, 2.6 inches the second day, 3.75 inches the third day, and 3 inches the fourth day. What was the average daily rainfall on Pedro's farm during those four days?

________

5. Jamie is taking five classes this semester. He has an 87 in English, an 82 in math, a 78 in science, a 91 in history, and a 96 in keyboarding. What is Jamie's overall average this semester? ________

*Check your work. Record your score.*

## ACTIVITY 6-4 YOUR GOAL: Get 12 of 15 answers correct.

Find the Rounded Problems. Use them to get the Estimated Answers. Then find the Calculator Answers. Write your answers in the blank spaces.

| | Original Problem | Rounded Problem | Estimated Answer | Calculator Answer |
|---|---|---|---|---|
| ● | 22 × 3.69 × .75 | 20 × 4 × .8 | 64 | 60.885 |
| 1. | 5.7 + 2.8 + 3.2 ÷ 3 | | | |
| 2. | 542 + 64 − 35 − 87 | | | |
| 3. | 8 + 2 ÷ 2.50 | | | |
| 4. | 16.95 × .07 + 16.95 | | | |
| 5. | 87 × 4 + 100 ÷ 5 | | | |

*Check your work. Record your score.*

# MAKING IT WORK

## REVIEW 6-1 YOUR GOAL: Get all answers correct.

### Business Forms

Combined operations are often used for calculations on business forms.

### Checkbook Register

A checking account lets a person make payments by check from money deposited in a bank. Each time a check is written, the information is recorded in a checkbook register. The person can then look at the checkbook register to find out how much money he or she has in the bank.

To work the checkbook register below:

1. Enter the beginning balance ($342.61) into your calculator.
2. Subtract the withdrawal ($250). The balance in the answer window will be $92.61.
3. Add the deposit ($35) to the balance. The answer ($127.61) should be in the answer window.
4. Find the ending balance by subtracting the withdrawals and adding the deposits for the remaining dates.

| Date | Description | Deposit | Withdrawal | Balance |
|---|---|---|---|---|
| 12/1 | Beginning Balance | | | $342.61 |
| 12/3 | Pam Gared (Rent) | | $250 | $92.61 |
| 12/10 | Kinu Oswan | $35 | | $127.61 |
| 12/20 | Food Day Groceries | | $75.56 | |
| 12/29 | Gears (paycheck) | $335.50 | | |
| 12/30 | Everelec (elec. bill) | | $27.86 | |

What is the ending balance? ________

## REVIEW 6-2 YOUR GOAL: Get all answers correct.

### Time Record

An Attendant's Time Record at Physicare Medical Center is shown on page 72. To complete the time record:

1. Subtract the earlier time (7:30) from the later time (12:30). Write the answer (5) in the Hours column.

2. Find the Hours for each day by repeating step 1 for each line.

3. Add the numbers in the Hours column. Write the answer in the space marked Total Hours.

4. Multiply the Total Hours by the Hourly Rate. Write the answer in the space marked Gross Salary.

PHYSICARE MEDICAL CENTER

Attendant's Time Record

*Patient*

Name Julia McCentric

Room # 534 Phone (605) 555-7239

*Attendant*

Name Kelly Carter

Address Blackwood Village, Apt. #507

Phone (605) 555-8215 Social Security # 346-58-7372

| Date of Service | Hours Worked | Hours |
|---|---|---|
| 7/3 | 7:30 – 12:30 | |
| 7/6 | 7:30 – 10:30 | |
| 7/9 | 8:00 – 12:00 | |
| 7/12 | 7:30 – 11:30 | |
| 7/15 | 7:00 – 12:00 | |

Total Hours ______

Hourly Rate $6.00

Gross Salary ______

Signature ______

Approved by Administrator ______

# UNIT 7

## MEMORY OPERATIONS—USING THEM EVERY DAY

### WHAT YOU WILL LEARN

When you finish this unit, you will be able to:

- Solve number problems using memory.
- Solve word problems using memory.

## THE BASICS OF MEMORY

The memory keys on your calculator let you store numbers to use later. The calculator "remembers" the numbers that you enter. Illustration 7-1 shows the memory keys labeled: $\boxed{M+}$ Memory-Plus, $\boxed{M-}$ Memory-Minus, and $\boxed{MRC}$ or $\boxed{M^{R}_{C}}$ Memory-Recall and Memory-Clear. Find the memory keys on your calculator.

Numbers can be put into the memory to be stored, added, or subtracted. The $\boxed{M^{R}_{C}}$ Key is called the **Memory-Recall and Memory-Clear** Key. This key lets you bring a number stored in

Illustration 7-1

Memory keys let you store numbers to use later.

memory back into the answer window by pressing it once. This stored number stays in memory until you clear the calculator's memory by pressing the [MRC] Key twice. If your calculator has an [AC] or **All Clear** Key, you can clear the memory and the answer window at the same time by pressing it once.

## Memory-Plus Key

The **Memory-Plus** Key [M+] adds a number to the memory. For example:

1. Clear the memory by pressing [MRC] two times.
2. Enter the number 847 into your calculator.
3. Press [M+]. An "M" will appear in the answer window of your calculator to show that the calculator has stored the number in memory.
4. Enter the number 649 into your calculator.
5. Press [M+].
6. Press [MRC]. 1,496 should appear in the answer window.

## Memory-Minus Key

The [M−] **Memory-Minus** Key subtracts a number from the memory. For example:

1. Clear the memory.
2. Work this problem: $3.65 + $2.18 + $1.15 =.
3. Press [M+] to add the total (6.98) to the calculator's memory.
4. Work this problem: $2.55 + $1.18 + $.95 = .
5. Press [M−] to subtract this total (4.68) from the total already in the memory.
6. Press [MRC] to get the difference between the two totals. The answer window should show 2.3.

## Multiplication and Memory

Memory can be used to add the products of multiplication problems. For example:

1. Clear the memory.
2. Find the answer to this problem: 25.67 × 9.
3. Press [M+].
4. Find the answer to this problem: 17.34 × 6.
5. Press [M+].
6. Press [MRC] to find the total (335.07) of the two products.

## CHECKPOINT 7-1

**YOUR GOAL:** Get 4 of 5 answers correct.

To work problems 1 through 5:

1. Clear your calculator and the memory.
2. Total the addends for the sample problem and compare your sum with the sum below the problem.
3. Add the sum to the memory by pressing $\boxed{M+}$ .
4. Repeat steps 2 and 3 for each group of numbers and record your answer below the problem.
5. Press $\boxed{M^{R}_{C}}$ to find the Grand Total.

| ● | | 1. | | 2. | | 3. | | 4. | | 5. |
|---|---|---|---|---|---|---|---|---|---|---|
| | 318 | | 977 | | 933 | | 197 | | 962 | GRAND |
| | 277 | | 380 | | 374 | | 902 | | 821 | TOTAL |
| | + 519 | | + 742 | | + 682 | | + 512 | | + 387 | |
| | 1114 | + | ______ | + | ______ | + | ______ | + | ______ | = ______ |

☞ ***Check your work. Record your score.***

### YESTERDAY, TODAY, AND TOMORROW

Bar codes are read by optical scanners which are electronic "eyes" in scanning plates or hand-held wands. The scanning plate that is built into the checkout counter is star-shaped, and has a clear covering. (See photo at right.) The hand-held wand is round and a little larger than a pencil. It has a clear covering at one end and a wire at the other end attached to a computer. The codes are made up of bars that stand for numbers. For the optical scanner to "read" the code, an item such as a loaf of bread or jar of jam is moved across a scanning plate. Or a hand-held wand can be moved across the item's bar code. The electronic "eye" enters the item name, quantity, and price into the computer. The computer prints the price and the name of the item on the customer's receipt and takes away the item from the store's inventory record.

# SOLVING WORD PROBLEMS—MEMORY

You can use your calculator's memory to solve word problems. Write down the key words and numbers in word problems. Then use these facts to solve the problem.

For example:

Bob wants to know how many cans of peaches and pears are in the storeroom. There are 12 cans in each box of peaches and 16 cans in each box of pears. There are 23 boxes of peaches and 17 boxes of pears. How many cans of fruit are there in the storeroom?

| | |
|---|---|
| Boxes of peaches: | 23 |
| Cans in each box: | 12 |
| Boxes of pears: | 17 |
| Cans in each box: | 16 |

To solve the problem:

1. Clear the memory.
2. Find the number of cans of peaches (23 × 12).
3. Store this answer (276) in the calculator's memory by pressing [M+] .
4. Find the number of cans of pears (17 × 16).
5. Add this answer (272) to the first answer by pressing [M+] .
6. Press [$M^R_C$] . The answer, 548, should appear in the answer window.

## Combining Memory Operations

Some memory problems can be solved by combining operations. For example:

Patty counted her tips after Roger's Restaurant closed. She counted:

13 - $1 bills
19 - quarters
11 - dimes
28 - nickels

How much did she earn in tips?

To work this problem:

1. Clear the memory.
2. Multiply the first two amounts (13 × 1).
3. Enter the product into the calculator's memory with [M+] .
4. Multiply the second set of factors (19 × .25). Enter the product into the memory with [M+] .
5. To make sure this number was added to the first answer, press [$M^R_C$] . The total, 17.75, should appear in the answer window.

6. Follow steps 1–5 for the remaining factors.
7. The answer should be $20.25. If you did not get $20.25, clear the calculator's memory. Work the problem again.

Remember to clear the calculator's memory before continuing.

## CHECKPOINT 7-2

**YOUR GOAL:** Get 4 of 5 answers correct.

If necessary, write the key words and numbers on a piece of paper. Use the memory keys to solve the problems. Write your answers in the blank spaces.

- Jill counted her tips after Billy's Restaurant closed. She counted:

  17 - $1 bills
  27 - quarters
  14 - dimes
  7 - nickels

  How much did she earn in tips? **$25.50**

1. Marla's part-time job is painting. She painted the Browning's apartment for $180. She worked four hours Monday, three hours Tuesday, and five hours Wednesday. How much did she get paid per hour? (*Hint:* 1. Total the amount of hours Marla worked. 2. Press [M+]. 3. Enter $180 into your calculator. 4. Press [÷]. 5. Press [M R/C]. 6. Press [=].)

   ______

2. Shane and Dana Thompson both work. Dana earns $4.25 per hour. Shane earns $7.60 per hour. Dana worked 37.5 hours last week. Shane worked 40 hours last week. How much did Shane and Dana earn all together? Round the answer to two decimals.

   ______

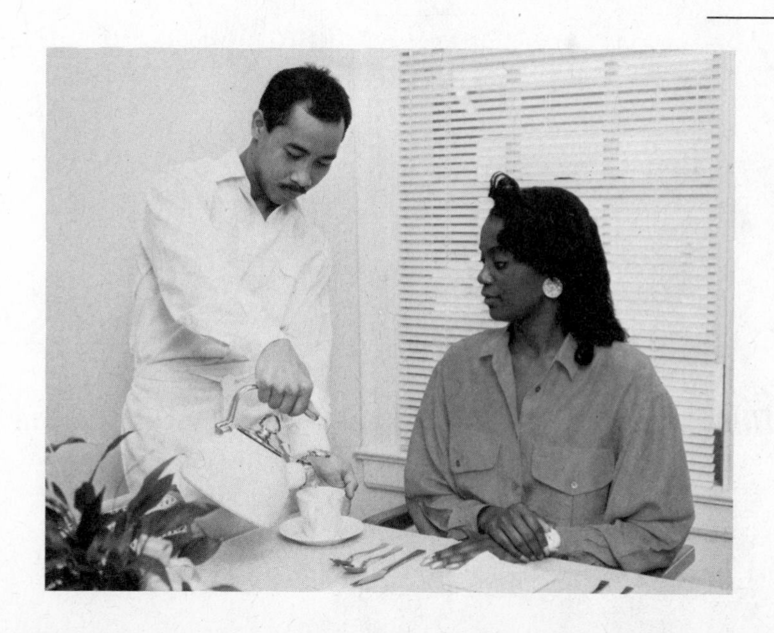

**3.** Mary needs to give her car a tuneup. She is going to buy a tune-up kit for \$7.99. She also needs four spark plugs and two quarts of oil. The spark plugs cost \$.89 each. The oil costs \$1.27 per quart. How much will she spend in all to tune up her car?

_______

**4.** Which brand of pants costs less? How much less?

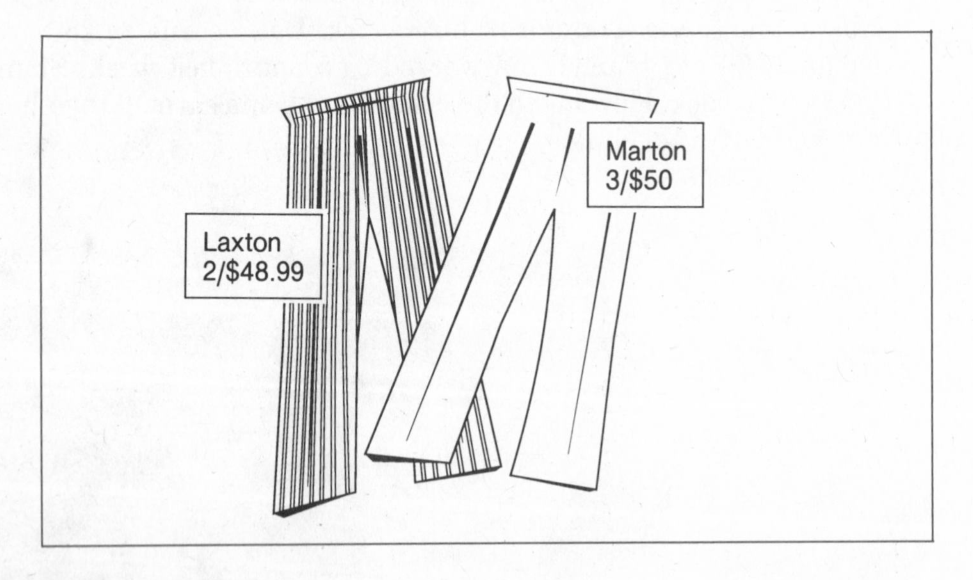


*Hint:* 1. Figure the cost for each pair of pants beginning with Laxton.
2. Compare the two prices.

_______ _______

**5.** At the grocery store, Jean finds that a brand-name shampoo costs $1.98 for 25 ounces. A generic brand costs $0.97 for 20 ounces. How much less is the generic brand per ounce? (*Hint:* 1. Figure the cost of the shampoo per ounce. 2. Compare the two prices.)

______

 ***Check your work. Record your score.***

## JUST FOR FUN

1. Pick a number and reverse it. (For example, if you pick the number 235, reverse it so it becomes 532.)
2. Subtract the smaller number from the larger number.

$$\begin{array}{r} 532 \\ -\ 235 \\ \hline 297 \end{array}$$

3. Reverse this number. (297 becomes 792.)
4. Add these two numbers.

$$\begin{array}{r} 297 \\ +\ \ 792 \\ \hline 1{,}089 \end{array}$$

This game can be played with any number and the answer will always be 1,089. Try it with another number.

## WHAT YOU HAVE LEARNED

Now that you have completed this unit, you should be able to:

- Solve number problems using memory.
- Solve word problems using memory.

# PUTTING IT TOGETHER

## ACTIVITY 7-1 YOUR GOAL: Get 4 of 5 answers correct.

To work problems 1 through 5:

1. Clear your calculator and the memory.
2. Total the addends for the sample problem and compare your sum with the sum below the problem.
3. Add the sum to the memory by pressing [M+] .
4. Repeat steps 2 and 3 for each group of numbers and record your answer below the problem.
5. Press [$M^R_C$] to find the Grand Total.

| ● | 1. | 2. | 3. | 4. | 5. GRAND TOTAL |
|---|---|---|---|---|---|
| 627 | 499 | 989 | 655 | 316 | |
| 336 | 734 | 663 | 321 | 823 | |
| + 874 | + 707 | + 853 | + 540 | + 142 | |
| 1837 + | ______ + | ______ + | ______ + | ______ = | ______ |

☞ *Check your work. Record your score.*

## ACTIVITY 7-2 YOUR GOAL: Get 4 of 5 answers correct.

Use the memory keys to solve the problems. Write your answers in the blank spaces.

● Anna wants to know how many servings of meat are in the storeroom. She counts four boxes of steaks and seven boxes of chicken. Each box of steak holds 36 steaks. Each box of chicken holds 25 chicken patties. How many servings of meat are in the storeroom?

319

1. Ramon needs new spark plugs and air hoses on his car. He replaces four spark plugs and two air hoses. Spark plugs cost $.98 each. Air hoses cost $11.95 each. How much will Ramon pay for the parts?

______

**2.** Andrea closed at Ernie's Grill last night. She counted 6 - $20 bills; 9 - $5 bills; 14 - $1 bills; 34 - $.25 coins; 12 - $.10 coins; 17 - $.05 coins; and 7 - $.01 coins. How much money was in the cash register?

________

**3.** Which brand of corn is cheaper per 8 oz. can? How much cheaper?

________ ________

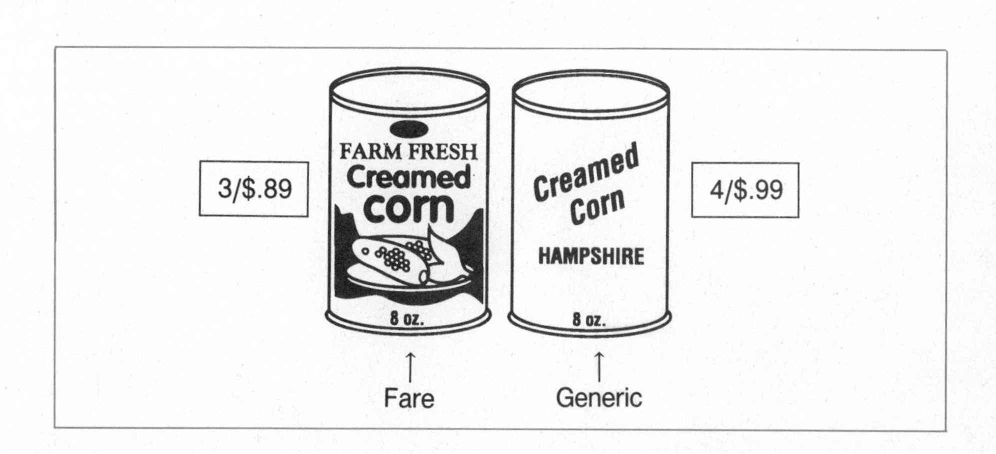


**4.** Sarita earned $256 before deductions last week. The Social Security deduction is $19.23. The Federal Income Tax deduction is $23. How much money did Sarita earn after her deductions?

________

**5.** Kelvin is running low on grain to feed his cattle. He had 21,960 pounds. For the last four days, he fed his cattle 3,660 pounds of grain each day. How much grain does Kelvin have left?

*Hint:* 1. Store the amount of grain he had before he fed his cattle (21,960) in memory.
2. Multiply the amount he fed his cattle each day (3,660) by the amount of days he has fed his cattle (4).
3. Press [M–] .
4. Press [MRC] to get your answer.

__________

☞ ***Check your work. Record your score.***

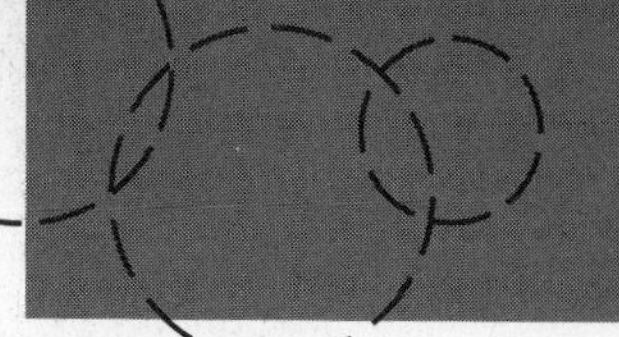

# MAKING IT WORK

## REVIEW 7-1 YOUR GOAL: Get all answers correct.

### Purchase Order

Below is a purchase order for Apple Appliances. To find the amount due:

1. Multiply the Quantity (2) by the Price Each ($298.95).
2. The amount due for the refrigerators ($597.90) should be in the answer window. Record your answer in the Amount Due column. Store this number in memory.
3. Find the amount due for each item by following steps 1 and 2.
4. The subtotal will be the sum of all of the amounts due. Recall the subtotal [M R/C] and record it in the Subtotal space. Multiply by 0.075 to find out how much sales tax will be added to the Subtotal. Record your answer in the Tax space. Then add the amount of the tax into memory [M+].
5. Recall the Total from memory [M R/C]. Record your answer in the Total space.

**Apple Appliances**

Purchase Order # 77761

| Item Number | Quantity | Description | Price Each | Amount Due |
|---|---|---|---|---|
| AQ477 | 2 | Refrigerator | $298.95 | ______ |
| AP3646 | 6 | Food Processor | 125.98 | ______ |
| QDAP-13 | 7 | Electric Knife | 21.95 | ______ |
| CPQ-62 | 3 | Blender | 63.92 | ______ |
| AP-3615 | 5 | Toaster | 25.95 | ______ |
| CPQ-65 | 4 | Hand Mixer | 17.85 | ______ |
| | | | Subtotal | ______ |
| | | | Tax (.075) | ______ |
| | | | Total | ______ |

Authorized by: ______________________

Ai-lien Stewart, Supplies Manager

6. Clear the memory and the answer window by pressing [AC]. (If your calculator does not have an [AC] key, press [M R/C] twice.)

☞ ***Check your work. Record your score.***

## REVIEW 7-2 YOUR GOAL: Get all answers correct.

## Inventory Form

To complete the inventory form:

1. Multiply the Number of Items under Stock Number 614 by the Unit Price ($4.99) to get the Inventory Price ($124.75). Press [M+].

THE TOY RACK
Inventory

Department/Shelf Vehicles/Shelf 9 Date 9/4/--

Counted by S. Hudson Priced by R. Cartwright

Recorded by M. Taylor

| Stock No. | Number of Items | Description | Unit Price | Inventory Price |
|---|---|---|---|---|
| 614 | | Fire Trucks | | |
| | 25 | red | 4.99 | 124.75 |
| | 10 | yellow | 4.99 | 49.90 |
| | 21 | white | 4.99 | 104.79 |
| | | | Total | 279.44 |
| 615 | | Tow Trucks | | |
| | 17 | blue and white | 3.49 | |
| | 16 | yellow | 3.49 | |
| | 19 | red | 3.49 | |
| | | | Total | |
| 616 | | Race Cars | | |
| | 31 | red | .89 | |
| | 31 | black | .89 | |
| | 31 | white | .89 | |
| | | | Total | |
| 617 | | Police Cars | | |
| | 20 | blue and white | 1.39 | |
| | 25 | black and white | 1.39 | |
| | | | Total | |
| | | Total Inventory Value | | |

**2.** Repeat step 1 for all of the items under Stock Number 614.

**3.** Press $\boxed{M^R_C}$ to get the total for Stock Number 614. Press $\boxed{M^R_C}$ again to clear your calculator.

**4.** Repeat steps 1–4 for Stock Numbers 615, 616, 617.

**5.** Add the Inventory Price Totals for each Stock Number. Record the Total Inventory Value for Shelf 9 at the bottom of the form.

☞ ***Check your work. Record your score.***

# PART THREE
## FRACTIONS, DECIMALS, AND PERCENTS

### UNIT 8
### FRACTIONS AND DECIMALS—USING THEM EVERY DAY

### UNIT 9
### PERCENTS—USING THEM EVERY DAY

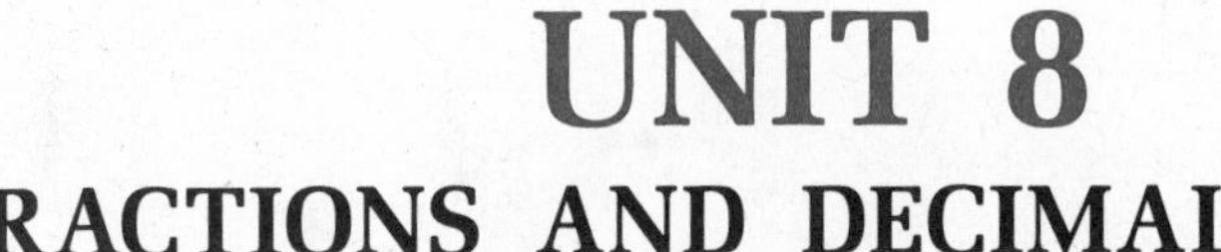

# UNIT 8

# FRACTIONS AND DECIMALS—USING THEM EVERY DAY

## WHAT YOU WILL LEARN

When you finish this unit, you will be able to:

- Identify different types of fractions.
- Change fractions to decimals.
- Solve fraction word problems.

## THE BASICS OF FRACTIONS

**Fractions** are whole numbers separated into equal parts. A fraction is written with a number above a line and a number below a line. The number above the line is called a **numerator**. The number below the line is called a **denominator**. The denominator shows how many parts the whole is separated into. The numerator shows how many parts are being talked about. Illustration 8-1 is an example of a fraction.

Illustration 8-1

A fractional number.

$\frac{2}{3}$ ← Numerator / ← Denominator

A pie graph may help you to understand what a fraction tells you. Illustration 8-2 is a pie graph that shows what $\frac{2}{3}$ of a whole is. The first picture shows you the three equal parts in the denominator. The second picture shows you the two parts you are talking about.

Illustration 8-2

A pie graph showing $\frac{2}{3}$.

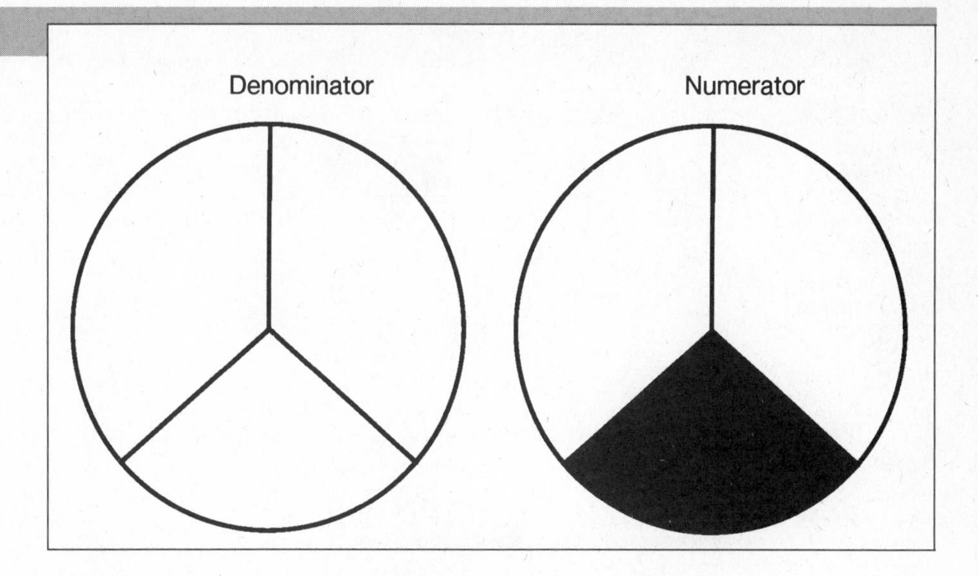


## TYPES OF FRACTIONS

There are several types of fractions. In a **proper fraction** the numerator is less than the denominator ($\frac{3}{4}$). In an **improper fraction** the numerator is greater than the denominator ($\frac{5}{2}$). A **mixed number** shows both a whole number and a fraction ($1\frac{6}{7}$).

Fractions can show how full a pan is of boiling water. For example, the pan is $\frac{1}{2}$ full. Proper fractions are used in recipes to tell how much of an ingredient (content in a mixture) to use. For example, add $\frac{2}{3}$ cup of water to a mix for brownies.

Illustration 8-3

## CHECKPOINT 8-1

**YOUR GOAL:** Get 4 of 5 answers correct.

Label each of the fractions as proper, improper, or mixed.

- ● $\frac{3}{4}$ **proper**
- **1.** $\frac{7}{8}$ ____________
- **2.** $2\frac{3}{5}$ ____________
- **3.** $\frac{13}{2}$ ____________
- **4.** $\frac{27}{51}$ ____________
- **5.** $69\frac{2}{3}$ ____________

☞ ***Check your work. Record your score.***

## READING FRACTIONS

Fractions can be read three different ways. For example, you could read $\frac{3}{4}$ as "three fourths," "three quarters," "three divided by four," or "three over four." The line that separates the numerator and the denominator means to divide. So $\frac{3}{4}$ could also be $3 \div 4$.

### JUST FOR FUN

In this game the answer will always be 23. Try it and see for yourself!

1. Pick any number.
2. Add 25 to the number.
3. Multiply the new number by 2.
4. Subtract 4 from this number.
5. Divide the number by 2.
6. Subtract the original number from this total.

The answer is 23. Try it with some other numbers.

## CHANGING FRACTIONS TO DECIMALS

Look at the keyboard of your calculator. There are no fractions on it. To work a problem with fractions on a calculator you must first change the fraction to a decimal. There are three ways to do this.

1. The fastest way is to memorize the decimal equivalents (same as or equal to) of fractions. For example, the decimal equivalent of $\frac{1}{2}$ is .5 and $\frac{1}{4}$ is .25.
2. The second fastest way is to look at a chart like this one:

| | | |
|---|---|---|
| $\frac{1}{2} = .5$ | $\frac{1}{4} = .25$ | $\frac{1}{5} = .2$ |
| $\frac{1}{3} = .333$ | $\frac{2}{4} = .5$ | $\frac{2}{5} = .4$ |
| $\frac{2}{3} = .667$ | $\frac{3}{4} = .75$ | $\frac{3}{5} = .6$ |
| | | $\frac{4}{5} = .8$ |

If you can memorize these fraction equivalents, you will not have to use a chart.

3. If there is not a chart or a fraction is not on the chart, use a calculator to change a fraction to a decimal.

For the fraction $\frac{1}{8}$, divide the numerator (1) by the denominator (8). The answer is .125.

Some fractions equal a decimal that repeats the same number. For example, use your calculator to change $\frac{2}{3}$ to a decimal. The window will show .66666. The sixes keep repeating. We can round decimals to any number of places. Round the decimal (.66666) to three places. The answer is now $\frac{2}{3}$ = .667.

## CHECKPOINT 8-2

**YOUR GOAL:** Get 4 of 5 answers correct.

Change each fraction to a decimal. Round the decimals to three places.

● $\frac{3}{7}$ = <u>.429</u>

1. $\frac{7}{8}$ = ______
2. $\frac{13}{28}$ = ______
3. $\frac{49}{56}$ = ______
4. $\frac{105}{213}$ = ______
5. $\frac{492}{625}$ = ______

***Check your work. Record your score.***

## SOLVING WORD PROBLEMS—FRACTIONS AND DECIMALS

Using a calculator to solve word problems with fractions is just a little different from solving word problems with whole numbers. You must first change the fractions or mixed numbers to decimals. Then you can work the problem.

Marty sees an ad in the newspaper that says "$\frac{1}{3}$ off the full price" at a discount store. Marty goes to the discount store and finds a pair of tennis shoes for $24. What is the sale price for the tennis shoes?

The answer to this word problem is easy to get. Follow these steps:

1. Divide 1 by 3 to get the decimal equivalent if you don't have a chart or have not memorized the equivalents.
2. Multiply 24 by .333 to get the amount of discount.
3. Enter the full price of $24 and subtract the discount to get the sale price.

**Memory Key Tips**

1. Enter 24 into memory.
2. Multiply 24 by 1 and divide by 3.
3. Press [M−].
3. Press [$M^{R}_{C}$].

Some word problems may have *fractions* that need to be added, subtracted, multiplied, or divided. Other problems will have *fractions and decimals* that need to be added, subtracted, multiplied, or divided. Change the fractions to decimals then work the problem.

Pam needs $1\frac{1}{4}$ of a pound of apples to make candied apples. She also needs $1\frac{1}{2}$ of a pound of apples for her family to eat. How many pounds of apples does she need all together?

1. Set up the problem. ($1\frac{1}{4} + 1\frac{1}{2} = ?$)
2. Change the fractions to decimals. ($1.25 + 1.50 = ?$)
3. Find the answer. (1.75)

Since $1.75 = 1\frac{3}{4}$, Pam will need $1\frac{3}{4}$ pounds of apples.

Illustration 8-4

## CHECKPOINT 8-3

**YOUR GOAL:** Get 4 of 5 answers correct.

Find the answer to each problem. Write your answer in the blank spaces.

- Karen needs $4\frac{1}{2}$ yards of material to make a dress for her daughter. She has picked out $2\frac{3}{4}$ yards. How many more yards of material does she need?

| | Need | | Have | | Answer |
|---|---|---|---|---|---|
| Fractions ⟶ | $4\frac{1}{2}$ | $-$ | $2\frac{3}{4}$ | $=$ | ? |
| Decimals ⟶ | 4.5 | $-$ | 2.75 | $=$ | 1.75 |
| Answer ⟶ | $1\frac{3}{4}$ yards | | | | |

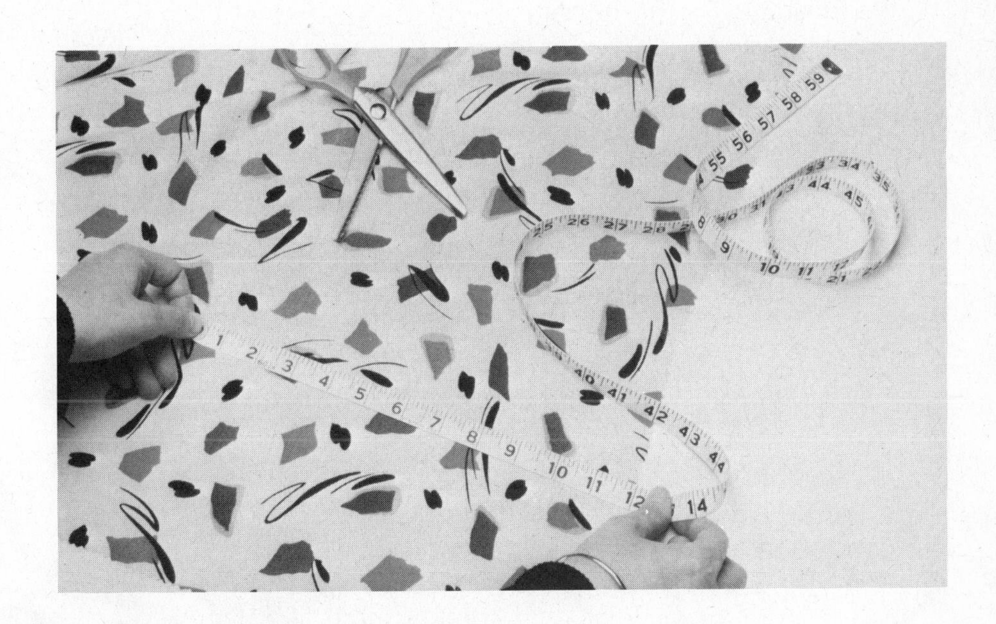

**1.** Stacey wants to make a pumpkin pie for Thanksgiving dessert. One of the ingredients for the recipe is $2\frac{3}{4}$ cups of flour. Most of her family will be at the dinner. She needs to make three pies to serve all of them. How much flour will she need?

________

**2.** Randy gets $280 per week for 40 hours of work. He had to leave town one afternoon for four hours and was paid $\frac{36}{40}$ of his salary. How much did he make that week?

________

**3.** Santa's Workshop hires people to work during the Christmas season. Joyce decided to work there to earn extra money for Christmas presents for her children. She made \$3.85 an hour and worked for $74\frac{1}{4}$ hours. How much money did she earn all together?

________

**4.** Greg needed to buy a long-sleeve shirt for the winter. He saw a "$\frac{1}{2}$ off" sale sign on a rack with long-sleeve shirts on it. Greg picked out a shirt from the rack that was \$15. How much was the sale price?

________

**5.** Casey's time record for one week showed she worked:

| | |
|---|---|
| Monday | $4\frac{1}{2}$ hours |
| Tuesday | $6\frac{1}{4}$ hours |
| Wednesday | $5\frac{1}{2}$ hours |
| Thursday | $7\frac{3}{4}$ hours |
| Friday | $3\frac{1}{4}$ hours |

How many total hours did she work that week?

________

☞ ***Check your work. Record your score.***

## YESTERDAY, TODAY, AND TOMORROW

Some hand-held calculators can do more than find answers to math problems. Some have clocks with alarms and stopwatches. Some play music.

## WHAT YOU HAVE LEARNED

Now that you have completed this unit, you should be able to:

- Correctly identify different types of fractions.
- Change fractions to decimals.
- Solve fraction word problems.

# PUTTING IT TOGETHER

## ACTIVITY 8-1 **YOUR GOAL:** Get 4 of 5 answers correct.

Label each of the fractions below as proper, improper, or mixed.

- $\frac{1}{6}$ proper

1. $\frac{16}{29}$ ________
2. $\frac{193}{56}$ ________
3. $2\frac{5}{6}$ ________
4. $\frac{45}{18}$ ________
5. $69\frac{3}{4}$ ________

*Check your work. Record your score.*

## ACTIVITY 8-2 **YOUR GOAL:** Get 4 of 5 answers correct.

Change each fraction to a decimal. Round each decimal to three places.

- $\frac{3}{4}$ .75

1. $\frac{45}{62}$ ________
2. $\frac{52}{49}$ ________
3. $\frac{274}{435}$ ________
4. $\frac{635}{1000}$ ________
5. $\frac{771}{296}$ ________

*Check your work. Record your score.*

## ACTIVITY 8-3 **YOUR GOAL:** Get 5 of 6 answers correct.

Find the answer to each problem. Write your answer in the blank spaces.

- Rosa wants to buy three donuts for her children. The price for a dozen (12) donuts is $2. What is the price for each donut ($\frac{1}{12} \times 2$)? How much will she pay for three donuts?

  $\frac{1}{12} \times 2 = ?$

  $.083 \times 2 = ?$     **$.166 or $.17 for each donut**

  $.17 \times 3 = ?$     **$.51 for 3 donuts**

1. Dean wants to buy $\frac{1}{2}$ pound of sugarless candy for trick-or-treaters at Halloween. One pound of this candy is $2.65. How much will he pay for $\frac{1}{2}$ pound?

_______

2. Tomatoes are on sale at the Thrifty Grocery Store for $.69 per pound. Glenda needs two tomatoes to make a salad for her family. Two tomatoes weigh $\frac{3}{4}$ of a pound. How much will she pay for two tomatoes?

_______

3. One yard of striped material at The Best Priced Cloth Store is $1.85. How much would $\frac{1}{4}$ of a yard of this material cost?

_______

4. Larry needs a coat for the winter. He found a coat for $50 on sale. The store requires $\frac{1}{10}$ of the sale price to put it in layaway. How much will Larry need to put the coat in layaway?

_______

5. Greg works at a gas station. He records the amounts of gas he pumps daily. Look at the Daily Gallons Pumped form. Follow the instructions to record the totals:

   1. Add all of the numbers together in the Regular and No-Lead columns for 7:00–8:00.
   2. Record your answer in the Totals column for 7:00–8:00.
   3. Follow steps 1 and 2 for the 8:00–9:00 row.

DAILY GALLONS PUMPED

| *Time* | *Regular* | *No-Lead* | *Totals* |
|---|---|---|---|
| 7:00–8:00 | 4.9 8.6 5.6 | 14.2 10.5 4.2 7.8 | |
| 8:00–9:00 | 5.6 2.3 | 9.1 13.9 4.5 8.9 3.6 9.23 | |

☞ ***Check your work. Record your score.***

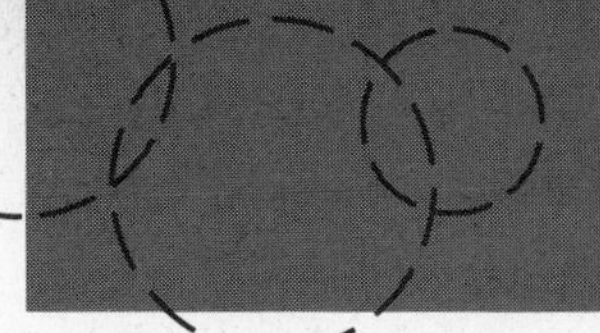

# MAKING IT WORK

## REVIEW 8-1 YOUR GOAL: Get all answers correct.

### Invoice

The invoice shown below is Jacksonville General Hospital's bill for desserts ordered for a birthday party from Yum Yum Bakery. To fill in the Amount column:

1. Change the mixed numbers in the Quantity column to decimals.
2. Multiply the Quantity by the Price. Write your answer in the Amount column.
3. Repeat steps 1 and 2 for all of the items on the invoice.
4. Add all of the amounts you recorded in the Amount column to get the Total of the invoice.

**YUM YUM BAKERY**
**5142 45th Street**
**Jacksonville, TX 79401**

**INVOICE NO.** 3456

**TO:** Jacksonville General Hospital
5013 19th
Jacksonville, TX 79401

**YOUR ORDER NO.** 538
**DATE** 1/19/93

| QUANTITY | DESCRIPTION | PRICE | AMOUNT |
|---|---|---|---|
| $2\frac{1}{2}$ dozen | Chocolate Chip Cookies | $1.19 per dozen | $2.98 |
| 2 | Birthday cakes | $16.73 | |
| $3\frac{1}{4}$ dozen | Peanut butter cookies | $.98 per dozen | |
| | | TOTAL | $ ________ |

Some Memory Key Tips are shown on page 98.

## Memory Key Tips

1. Change the mixed numbers in the Quantity column to decimals.
2. Multiply the quantity by the Price. Write your answer in the Amount column. Press [M+] .
3. Repeat steps 1 and 2 for all of the items on the invoice.
4. Press [M R/C] .

***Check your work. Record your score.***

## REVIEW 8-2 YOUR GOAL: Get all answers correct.

### Sales Slip

Shown below is Janet's sales slip from Cost Less Cloth Store. To fill in blanks in the Amount column:

**1.** Change the fractions in the Quantity column to decimals.

**Cost Less Cloth Store**
**1616 34th Street**
**Orlando, FL 32809**

NO 8154

DATE 5/9/93

NAME Janet Holcomb

ADDRESS 5511 68th Street
Orlando, FL 32810

| Cash | Charge | C.O.D. | Send | Taken |
|---|---|---|---|---|
| ✓ | | | | |

| Qty | Description | Unit Price | Amount |
|---|---|---|---|
| 1/4 yd. | polka dot | 1.25 | |
| | | | |
| 1 3/8 yd. | striped | .89 | |
| | | | |
| 3/4 yd. | plain (red) | 2.59 | |
| | | | |
| | | | |
| | | TOTAL | |

2. Multiply the decimals by the amount in the Price column.
3. Round your answer to two places. Record your answer in the Amount column.
4. After you have completed the Amount column for all of the material, add all of the amounts to get the Total of the sale.

## Memory Key Tips

1. Change the fractions in the Quantity column to decimals.
2. Multiply the decimals by the amount in the Price column.
3. Round your answer to two places. Record your answer in the Amount column. Press [M+] .
4. After you have completed the Amount column for all of the material, press [$M^R_C$] .

☞ ***Check your work. Record your score.***

# UNIT 9
## PERCENTS—USING THEM EVERY DAY

**WHAT YOU WILL LEARN**

When you finish this unit, you will be able to:

- Change percents to decimals.
- Change decimals to percents.
- Find percentages.
- Find the percent of increase and decrease.

## THE BASICS OF PERCENTS

### What is a Percent?

A **percent** is another way to compare a part with a whole. You learned about comparing parts with wholes using fractions and decimals in Unit 1. Percents are shown by the symbol %. For example, 25%.

A dollar bill is an easy way to see how fractions, decimals, and percents are all ways of showing what part something is of a whole.

Illustration 9-1

The shaded portion of this dollar can represent a fraction, a decimal, or a percent.

The first part of the dollar can be written as a fraction, a decimal, or a percent.

| Parts of a Whole | Math Symbol | Name |
|---|---|---|
| Fraction | $\frac{1}{4}$ | One quarter |
| Decimal | .25 | 25¢ |
| Percent | 25% | 25 of 100 (per)cents |

## Changing Percents to Decimals

Many hand-held calculators have [%] Percent Keys to help you find the percent of a number. Some calculators may not have a [%] Percent Key. In this case, you will need to change a percent to a decimal.

To change a percent to a decimal, think of the two circles in the % sign as two decimal places in the number. Remove the % sign and move the decimal two places to the left.

1. Some percentages do not have a decimal written in the number (27%). A whole number can also be written with a decimal to the right (27% = 27.%). Move the decimal two places to the left and drop the percent sign. For example,

   27% = 27.% = .27

2. Some percentages have a decimal in the number (3.5%). Move the decimal two places to the left. You may have to add one or more zeros. Take away the % sign. For example,

   3.5% = 03.5% = .035

Changing a percent to a decimal is used to find how much an item will cost after a percentage discount. For example,

Sandy went to a 25% off sale at a clothing store. She found a pair of slacks for $28.99. How much will the slacks cost after the discount?

To work this problem if your calculator:

A. Has a [%] key:
   1. Enter the price of the slacks (28.99).
   2. Press [×]. Enter the percentage (25).
   3. Press [%]. The amount of the discount, $7.2475, will be in the window.
   4. Enter the price of the slacks (28.99). Press [−].
   5. Enter the amount of the discount (7.2475). Press [=].
   6. The answer, 21.74, should appear in the window.

B. Does not have a [%] key:

1. Enter the price of the slacks (28.99). Press [×].
2. Change the percent to a decimal (25% = .25).
3. Enter the decimal discount (.25) into your calculator.
4. Press [=]. The amount, 7.2475, should appear in the window.
5. Enter the price of the slacks (28.99). Press [−].
6. Enter the amount of the discount (7.2475). Press [=].
7. The answer, 21.74, should appear in the window.

**Memory Key Tips**

To work this problem if your calculator:

A. Has a [%] key:

1. Enter 28.99. Press [M+].
2. Multiply 28.99 by 25%.
3. Press [M−].
4. Press [MRC].

B. Does not have a [%] key:

1. Enter 28.99 Press [M+].
2. Multiply 28.99 by .25.
3. Press [M−].
4. Press [MRC].

## Changing Decimals to Percents

To change a decimal to a percent, move the decimal point to the right two places and put a percent sign after the number. For example, to change .35 to a percent:

.35 = .35. = 35%

## CHECKPOINT 9-1

**YOUR GOAL:** Get 9 of 10 answers correct.

The Modern Furniture Store had a Labor Day sale at discounts shown in the form. What is the sale price of each item?

To complete the form:

**1.** Enter the Original Price (34.15).

**2.** Change the Discount percentage (27%) to a decimal. Multiply the Original Price by the decimal discount (.27).

**3.** Record the Discount Amount (9.22) in the Discount Amount column.

**4.** Subtract the Discount Amount from the Original Price.

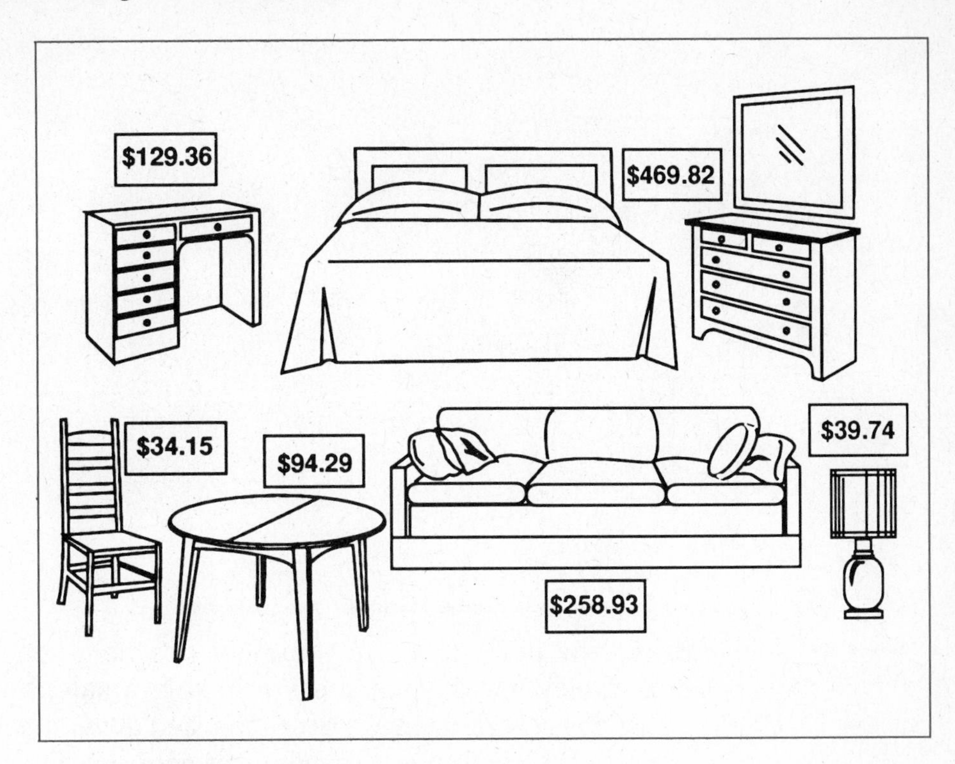


**5.** Record the Sale Price (24.93) in the Sale Price column.

**6.** Repeat steps 1–5 for problems 1–5.

### Memory Key Tips

1. Enter 34.15. Press [M+].
2. Multiply 34.15 by .27.
3. Press [M−].
4. Press [M R/C].

| Description | Discount % | Original Price | Discount Amount | Sale Price |
|---|---|---|---|---|
| Kitchen chair | 27% | $34.15 | ● 9.22 | ● $24.93 |
| Dining Table | 51% | $94.29 | 1. ______ | 2. ______ |
| Sofa | 29% | $258.93 | 3. ______ | 4. ______ |
| Lamp | 15.8% | $39.74 | 5. ______ | 6. ______ |
| Bedroom suite | 35.2% | $469.82 | 7. ______ | 8. ______ |
| Desk | 46.5% | $129.36 | 9. ______ | 10. ______ |

☞ ***Check your work. Record your score.***

## JUST FOR FUN

What always happens when you play the following game?

1. Enter any number on the calculator.
2. Multiply the number by 3.
3. Add 30 to this number.
4. Multiply the new number by 5.
5. Add 600 to your number.
6. Divide this number by 15.
7. Subtract 50 from the new number.

What always happens?

## FINDING THE PERCENTAGE

Finding the percentage of a number can help you to know how much money you save at a store having a sale. For example,

The Designer Furniture Store marked down most of their furniture for a big going-out-of-business sale. Jessica needed to buy a folding chair for her kitchen. The original price of the chair was $24.95 and was marked down to $18.95. What percentage would she save if she bought the chair? ________

To solve this problem:

1. Find the amount of money she will save (24.95 − 18.95 = 6.00).
2. Divide the amount she saves (6) by the original price of the chair ($24.95). (6 ÷ 24.95 = .2405).
3. Change the decimal in your answer (.2405) to a percent (24.05%).

## CHECKPOINT 9-2

YOUR GOAL: Get 4 of 5 answers correct.

Jake's Restaurant Supply Store pays their salespersons on a commission basis. This means that they are paid a percentage of their sales. The table shows the amounts of commission each salesperson makes.

To complete the table:

1. Divide the Commission Amount by the Sales.
2. Change your answer to a percent. Round your answer to one decimal place.
3. Record your answer in the Commission Percentage column.
4. Repeat steps 1, 2, and 3 for the rest of the salespersons.

JAKE'S RESTAURANT SUPPLY STORE

Salespersons' Commissions
for Week Ending: May 16

| **Salesperson** | **Sales** | **Commission Amount** | **Commission Percentage** |
|---|---|---|---|
| Jerry Smith | $895.28 | $ 85.95 | ● 9.6% |
| Carlota Jordan | $937.61 | $136.30 | **1.** ______ |
| Dan Johnson | $3,028.53 | $908.56 | **2.** ______ |
| Harvey Mattison | $1,420.69 | $247.20 | **3.** ______ |
| Jun Taylor | $2,573.48 | $635.65 | **4.** ______ |
| Cindy Carnwrick | $2,836.29 | $731.76 | **5.** ______ |

☞ ***Check your work. Record your score.***

## FINDING PERCENT OF INCREASE OR DECREASE

Differences between two amounts are often shown by percents to help a person better understand the differences.

Jeff's rent for his apartment increased from $275 to $300 a month. What is the percent of increase?

To find the percent of increase (or decrease):

1. Find the difference between the original amount (275) and the new amount (300). (300 − 275 = 25).
2. Divide the difference (25) by the original amount (275). (25 ÷ 275 = .091).
3. Change the answer to a percent. (.091 = 9.1%. There was a 9.1% increase.)

## *CHECKPOINT 9-3*

**YOUR GOAL:** Get 4 of 5 answers correct.

Find the percent of increase or decrease. Round your answers to two decimal places. Record your answers in the blank spaces.

● Heather was making $275 per week at her job. She received a raise and started getting $294.75 per week. What was the percent of increase?

294.75 − 275 = 19.75
19.75 ÷ 275 = .07182
.07182 = 7.18%

1. The price of hamburger meat rose from $1.27 per pound to $1.57 per pound. What was the percent of increase?

______

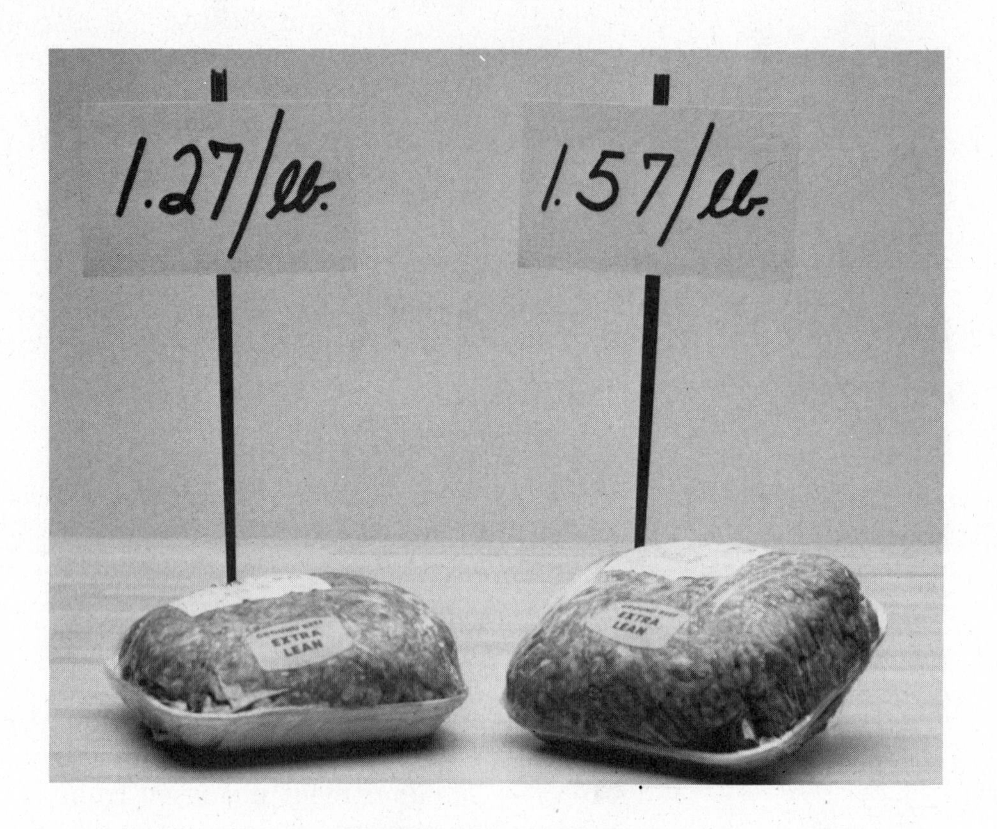


2. Ms. Mosier received an annual pay raise of $500. Her original salary was $18,256.71. What was the percent of increase?

______

3. Ken Farr bought a car for $28,500. He soon discovered that the payments were too high for him to make. He sold the car for $25,000. What was the percent of loss?

______

4. At a sale, Sally found a shirt for $25 at regular price reduced to $19.98. What was the percentage of decrease in the price?

______

5. Harry paid his electric bill of $62.71 last month. This month his bill increased to $67.10. What was the percentage of increase?

______

☞ ***Check your work. Record your score.***

## CALCULATORS MAKE IT EASY!

Calculators make it easy to shop while on a budget. When you take your calculator to the store, you can add the cost of items as you pick them up. This way you will be sure to have enough money to pay for all of the items in your cart.

This shopper is adding the cost of the items in his cart. He does not want the cost of the items in his cart to be more than the amount of money he has to spend.

## WHAT YOU HAVE LEARNED

Now that you have completed this unit, you should be able to:

- Change percents to decimals correctly.
- Change decimals to percents correctly.
- Find percentages.
- Find percents of increase or decrease.

# PUTTING IT TOGETHER

## ACTIVITY 9-1 YOUR GOAL: Get 4 of 5 answers correct.

The expenses (amount of money spent for supplies, salaries, etc.) for the Tot Toy Company totaled $118,240. Shown are the percentages of the total amount from each department in the corporation.

To complete the form if your calculator:

**A.** Has a [%] key:

1. Enter the total amount of expenses (118,240). Press [×].
2. Enter the Percent of Expenses (19.5). Press [%].
3. Round your answer to the nearest dollar. (If your answer includes $.50 or more, round the dollar up. If your answer is less than $.50, do not change the dollar amount. For example, $23.613 = $23.61 = $24 or $58.391 = $58.39 = $58.) Record your answer in the Amount of Expenses column.
4. Repeat steps 1–3 for the other departments.
5. Prove your work by adding the part (Amount of Expenses including Sales of $23,057) to get the whole (TOTAL).

**B.** Does not have a [%] key:

1. Enter the total amount of expenses (118,240). Press [×].
2. Change the percent to a decimal and enter the decimal amount into your calculator (.195).
3. Press [=]. Round your answer to the nearest dollar. (If your answer includes $.50 or more, round the dollar up. If your answer is less than $.50, do not change the dollar amount. For example, $23.613 = $23.61 = $24 or $58.391 = $58.39 = $58.) Record your answer in the Amount of Expenses column.
4. Repeat steps 1–3 for the other departments.
5. Prove your work by adding the part (Amount of Expenses including Sales of $23,057) to get the whole (TOTAL).

| Department | Percent of Expenses | Amount of Expenses |
|---|---|---|
| Sales | 19.5% | ● $23,057 |
| Accounting | 9.6% | 1. ______ |
| Warehouse | 12.8% | 2. ______ |
| Production | 32.7% | 3. ______ |
| Advertising | 14.9% | 4. ______ |
| Personnel | 10.5% | 5. ______ |
| TOTAL | | $118,240 |

***Check your work. Record your score.***

## ACTIVITY 9-2 YOUR GOAL: Get 9 of 10 answers correct.

Retail stores add a percent (30%) of the cost of an item ($85.34) to the cost to make a profit ($25.60). This percent is called the "markup." The cost plus the markup equals the selling price ($110.94).

This dress was bought for $85.34 and will be sold for $110.94.

| $85.34 | | $25.60 | | $110.94 |
|---|---|---|---|---|
| Cost | + | Markup | = | Selling Price |
| 100% | | 30% | | 130% |

If you know the cost and selling price and want to find the percent of markup, divide the amount of markup (25.60) by the cost price (85.34). (25.60 ÷ 85.34 = .30 = 30%).

To find the percent of markup, complete the form:

1. Subtract the Cost of the dress (85.34) from the Selling Price (110.94) of the dress (110.94 − 85.34 = 25.60). This is the markup.

2. Divide markup (25.60) by the Cost (85.34). The answer is the percent of markup (30%).

3. Round your answer to two decimal places. Record your answer in the Percent of Markup column.

CHARLOTTE'S CLOTHING

| Type of Dress | Cost | Selling Price | Amount of Markup | Percent of Markup |
|---|---|---|---|---|
| Evening gown | $85.34 | $110.94 | ● $25.60 | ● 30% |
| Party dress | 35.12 | 53.98 | 1. ______ | 2. ______ |
| Work suit | 57.91 | 69.35 | 3. ______ | 4. ______ |
| Casual dress | 28.49 | 34.99 | 5. ______ | 6. ______ |
| Black skirt | 14.23 | 19.97 | 7. ______ | 8. ______ |
| White skirt | 15.67 | 21.65 | 9. ______ | 10. ______ |

☞ ***Check your work. Record your score.***

## ACTIVITY 9-3 YOUR GOAL: Get 4 of 5 answers correct.

Find the percent of increase or decrease. Round your answers to two places. Record your answers in the blank spaces.

● John is going to buy some notebook paper for his night class that costs $1.29. The total price of the paper including sales tax was $1.37. What percent of sales tax was added to the original price of the paper?

$1.37 − $1.29 = $.08 $.08 ÷ $1.29 = .062 = 6.2%

1. Reka wants to buy a purse that originally costs $11.99. The final price of the purse is $12.89. How much sales tax was added to the price of the purse?

_______

2. Matt and Terry ate dinner at a restaurant. The total amount of their bill was $17.34. They left $20, which included the tip. What percent of the total bill was paid as their tip?

_______

3. Jake wants to sell you a used car for $2,100. You offer $300 less, or $1800. What percent will you save if Jake accepts your offer?

_______

4. Shannon was told by a friend that he should only spend 25% of his gross pay a month on rent for an apartment. Shannon makes $1,250 a month. How much can he spend on an apartment?

_______

5. George's Clothing Store bought black leather belts at a cost of $10.32 each. The selling price of each belt is $15.98. What is the percent of markup on the belts?

_______

***Check your work. Record your score.***

# PART FOUR
## BUSINESS FORMS SIMULATION

# PART 4

## BUSINESS FORMS SIMULATION

### WHAT YOU WILL LEARN

When you finish this part, you will be able to:

- Review calculator operations by completing a car repair order simulation. A **simulation** is practice working with forms as they are used in a real business.
- Calculate orders for a car repair shop.

## ABOUT THE CAR CLINIC

The Car Clinic is a repair shop in Los Angeles, California. You work for the Car Clinic and your jobs are:

1. Delivering and picking up cars for repair.
2. Cleaning up and putting tools away.
3. Getting parts.
4. Completing forms.

One day when you were not too busy, the owner, Mr. Horton, asks, "Do you know how to use a calculator?"

You reply, "Yes. I learned how to use one in the Adult Education course I took."

He says, "Great! Will you complete these repair orders for me?"

You gladly answer, "Sure."

Mechanics complete a repair order when cars are brought to the shop. See page 116 for an example of a repair order. A repair order is a form that gives details about the customer, the car, and repairs to the car.

## What You Are To Do

1. Clear your calculator's memory.
2. Calculate the Total Parts. In the Parts section of Repair Order 1, add the amounts for each part used. Write your answer at the bottom of the Parts section in the Total Parts space. Also write your answer in the Total Parts space in the Labor section. Press [M+] to add your total in the memory.
3. Calculate the Sales Tax. Multiply the Total Parts amount by the 7% sales tax. Use the [%] key if your calculator has one. Round your answer to two decimal places. Write your answer in the Tax on Parts space in the Labor section. (Tax is not charged on labor.) Press [M+] to add the tax to the total parts amount in memory.
4. Calculate the Total Labor. In the Labor section of the repair order, add the Amounts for the cost of labor. Write your answer in the Total Labor space. Press [M+] to add this total to the tax and total parts amounts.
5. Calculate the Total. Press [$\text{M}^{\text{R}}_{\text{C}}$] to get the total of the repair order. Round your answer to two decimal places. Write your answer in the TOTAL space.
6. Check your work. You want to make sure that your work is correct so that a customer is billed fairly. Check your answers on Repair Order 1 with the answers at the back of the book to be sure you are completing the form correctly.
7. Record your number of correct answers on the Personal Progress Record on page 141.
8. Repeat steps 1–7 for Repair Orders 2–12 on page 116–121.

## Checking Your Work

After you complete Repair Orders 2–12, check your work with the answers at the back of the book. Record the number of correct answers on the Personal Progress Record on page 141.

**1.**

| PARTS | | | | |
|---|---|---|---|---|
| Quantity | Number | Description | Amount | |
| 1 | H0171 | Head Set | 97 | 45 |
| 1 | H3661 | Smog Tube | 14 | 69 |
| 1 | H4003 | Heater Hose | 12 | 50 |
| 1 | F0232 | Oil Filter | 8 | 41 |
| 2 | A2770 | Gals. Antifreeze | 15 | 84 |
| 4 | C1587 | Rocker Clips | 4 | 50 |
| 5 | 018125 | Qts. Oil | 10 | 00 |
| | | Total Parts | | |

**CAR CLINIC**
**1257 Madison Street**
**Los Angeles, CA 90035-1257**

**REPAIR ORDER**

Name: Earl Scudday Date: 5/11/--
Address: 916 Maple St. Phone: 275-7711

| Make & Model | Year | License | Mileage |
|---|---|---|---|
| Chevy Silverado | 1989 | PCB-695 | 25,503 |

| Mechanic's Initials | LABOR Description | Amount | |
|---|---|---|---|
| RS | Replace heads and clean, replace gasket | 375 | 50 |
| WI | Change oil and filter add antifreeze | 15 | 75 |
| Comments: | TOTAL LABOR | | |
| | TOTAL PARTS | | |
| | 7% TAX ON PARTS | | |
| | TOTAL | | |

**2.**

| PARTS | | | | |
|---|---|---|---|---|
| Quantity | Number | Description | Amount | |
| 1 | M7625 | Window Motor | 107 | 43 |
| | | Total Parts | | |

**CAR CLINIC**
**1257 Madison Street**
**Los Angeles, CA 90035-1257**

**REPAIR ORDER**

Name: Kenneth Seaberg Date: 5/11/--
Address: 2132 Aberdeen Phone: 274-2391

| Make & Model | Year | License | Mileage |
|---|---|---|---|
| Volvo 240 DL | 1990 | RBP-703 | 13,113 |

| Mechanic's Initials | LABOR Description | Amount | |
|---|---|---|---|
| WI | Check and replace left motor | 37 | 50 |
| Comments: | TOTAL LABOR | | |
| | TOTAL PARTS | | |
| | 7% TAX ON PARTS | | |
| | TOTAL | | |

**3.**

| PARTS | | | | |
|---|---|---|---|---|
| Quantity | Number | Description | Amount | |
| 1 | B1920 | Set wheel bearing | 71 | 80 |
| 2 | A2770 | Gals. antifreeze | 15 | 84 |
| 2 | F8376 | Qts. brake fluid | 15 | 00 |
| | | Total Parts | | |

CAR CLINIC
1257 Madison Street
Los Angeles, CA 90035-1257

REPAIR ORDER

Name: Arnold Phelps Date: 5/11/--
Address: 6505 16 St. Phone: 275-7990

| Make & Model | Year | License | Mileage |
|---|---|---|---|
| Ford Astro Van | 1987 | FYD-696 | 57,273 |

| Mechanic's Initials | LABOR Description | Amount | |
|---|---|---|---|
| WI | Flush radiator | 21 | 50 |
| JR | Pack front and back wheel bearings - replace | 123 | 50 |
| JR | Drain and replace brake fluid | 17 | 75 |
| Comments: | TOTAL LABOR | | |
| | TOTAL PARTS | | |
| | 7% TAX ON PARTS | | |
| | TOTAL | | |

**4.**

| PARTS | | | | |
|---|---|---|---|---|
| Quantity | Number | Description | Amount | |
| 5 | 08125 | Qts. oil | 10 | 00 |
| 1 | F0232 | Oil Filter | 8 | 41 |
| 1 | A2770 | Gals. antifreeze | 7 | 92 |
| 4 | F8377 | Trans. fluid | 24 | 20 |
| | | Total Parts | | |

CAR CLINIC
1257 Madison Street
Los Angeles, CA 90035-1257

REPAIR ORDER

Name: Mr. Curtis Walker Date: 5/12/--
Address: 1415 25 St. Phone: 277-4849

| Make & Model | Year | License | Mileage |
|---|---|---|---|
| Mazda RX7 | 1988 | BAS-175 | 27,377 |

| Mechanic's Initials | LABOR Description | Amount | |
|---|---|---|---|
| JR | Change oil and filter | 27 | 50 |
| RS | Rotate tires | 10 | 00 |
| JR | Realign front end | 73 | 75 |
| RS | Change transmission fluid | 11 | 50 |
| RS | Check and refill fluid | 3 | 00 |
| Comments: | TOTAL LABOR | | |
| | TOTAL PARTS | | |
| | 7% TAX ON PARTS | | |
| | TOTAL | | |

**5.**

| PARTS | | | | |
|---|---|---|---|---|
| Quantity | Number | Description | Amount | |
| 1 | P5026 | Set disc pads | 61 | 25 |
| 1 | A5631 | Axle boot | 28 | 28 |
| 1 | B4387 | Set brake shoes | 21 | 50 |
| 1 | C7010 | Master cylinder | 51 | 56 |
| 1 | P0075 | Tire patch | | 79 |
| | | Total Parts | | |

**CAR CLINIC**
**1257 Madison Street**
**Los Angeles, CA 90035-1257**

**REPAIR ORDER**

Name: Maria Egbe Date: 5/12/--
Address: 3124 Utica Phone: 275-7638

| Make & Model | Year | License | Mileage |
|---|---|---|---|
| Chevy S-10 | 1989 | GLU-269 | 22,313 |

| Mechanic's Initials | LABOR Description | Amount | |
|---|---|---|---|
| TR | Realign front and rear | 64 | 00 |
| TR | Replace master cylinder | 34 | 75 |
| RS | Patch leak Rt. front tire | 3 | 00 |
| Comments: | TOTAL LABOR | | |
| | TOTAL PARTS | | |
| | 7% TAX ON PARTS | | |
| | TOTAL | | |

**6.**

| PARTS | | | | |
|---|---|---|---|---|
| Quantity | Number | Description | Amount | |
| 1 | B1283 | Bulb | | 89 |
| | | Total Parts | | |

**CAR CLINIC**
**1257 Madison Street**
**Los Angeles, CA 90035-1257**

**REPAIR ORDER**

Name: Sandy Hensley Date: 5/13/--
Address: Kenosha Dr. Phone: 275-4429

| Make & Model | Year | License | Mileage |
|---|---|---|---|
| Buick Park Avenue | 1989 | TEN-823 | 18,503 |

| Mechanic's Initials | LABOR Description | Amount | |
|---|---|---|---|
| RS | State Inspection | 8 | 00 |
| RS | Replace Rt. tail light | 2 | 00 |
| Comments: | TOTAL LABOR | | |
| | TOTAL PARTS | | |
| | 7% TAX ON PARTS | | |
| | TOTAL | | |

**7.**

**CAR CLINIC**
**1257 Madison Street**
**Los Angeles, CA 90035-1257**

**REPAIR ORDER**

Name: Brian Heflin  Date: 5/13/--
Address: 1914 38 St.  Phone: 275-4429

| Make & Model | Year | License | Mileage |
|---|---|---|---|
| Buick Regal | 1990 | QGA-187 | 11,012 |

| PARTS | | | | |
|---|---|---|---|---|
| Quantity | Number | Description | Amount | |
| 1 | T9915 | Steel-belted | | |
| | | radial tire | 65 | 72 |
| | | Total Parts | | |

| Mechanic's Initials | LABOR Description | Amount | |
|---|---|---|---|
| WI | State Inspection | 8 | 00 |
| WI | adjust left | | |
| | head light | 2 | 50 |
| JR | Replace worn | | |
| | rt. front tire | 3 | 00 |

| Comments: | | |
|---|---|---|
| TOTAL LABOR | | |
| TOTAL PARTS | | |
| 7% TAX ON PARTS | | |
| TOTAL | | |

**8.**

**CAR CLINIC**
**1257 Madison Street**
**Los Angeles, CA 90035-1257**

**REPAIR ORDER**

Name: Daniel Garcia  Date: 5/13/--
Address: 2634 Xavier  Phone: 276-6832

| Make & Model | Year | License | Mileage |
|---|---|---|---|
| Volkswagen Quantum | 1988 | YGN-980 | 34,877 |

| PARTS | | | | |
|---|---|---|---|---|
| Quantity | Number | Description | Amount | |
| 6 | 55729 | Spark plugs | 21 | 42 |
| 1 | D3865 | Distributor cap | 32 | 52 |
| 1 | W1039 | Set wires | 61 | 98 |
| 1 | R2001 | Rotor | 9 | 58 |
| 1 | F2155 | Fuel filter | 4 | 95 |
| 1 | F3284 | air filter | 12 | 19 |
| 1 | H9039 | Vacuum hose | 3 | 79 |
| | | Total Parts | | |

| Mechanic's Initials | LABOR Description | Amount | |
|---|---|---|---|
| RS | Tune engine | 41 | 25 |
| JR | Install distributor | | |
| | wires filters | 13 | 50 |

| Comments: | | |
|---|---|---|
| TOTAL LABOR | | |
| TOTAL PARTS | | |
| 7% TAX ON PARTS | | |
| TOTAL | | |

**9.**

| PARTS | | | | |
|---|---|---|---|---|
| Quantity | Number | Description | Amount | |
| 1 | P0197 | Water pump | 71 | 48 |
| 1 | B9606 | Alternator belt | 13 | 19 |
| | | Total Parts | | |

CAR CLINIC
1257 Madison Street
Los Angeles, CA 90035-1257

**REPAIR ORDER**

Name: Jolee's Restaurant Date: 5/14/--
Address: 2411 Slide Rd. Phone: 265-2001

| Make & Model | Year | License | Mileage |
|---|---|---|---|
| Plymouth Voyager | 1988 | LJW-491 | 33,786 |

| Mechanic's Initials | LABOR Description | Amount | |
|---|---|---|---|
| JR | Replace water pump and alternator belt | 46 | 88 |
| | Needs front springs | | |

| Comments: | | |
|---|---|---|
| TOTAL LABOR | | |
| TOTAL PARTS | | |
| 7% TAX ON PARTS | | |
| TOTAL | | |

**10.**

| PARTS | | | | |
|---|---|---|---|---|
| Quantity | Number | Description | Amount | |
| 2 | A2770 | Gals. antifreeze | 15 | 84 |
| 1 | T6626 | Thermostat and gasket | 12 | 32 |
| 1 | B8069 | Blower motor | 127 | 67 |
| 1 | F7787 | Fuse | | 27 |
| 1 | R1123 | Regulator | 33 | 59 |
| | | Total Parts | | |

CAR CLINIC
1257 Madison Street
Los Angeles, CA 90035-1257

**REPAIR ORDER**

Name: Greg Tidmore Date: 5/15/--
Address: 5728 75 St. Phone: 277-3610

| Make & Model | Year | License | Mileage |
|---|---|---|---|
| Pontiac Fiero | 1990 | JAD-328 | 11,008 |

| Mechanic's Initials | LABOR Description | Amount | |
|---|---|---|---|
| JR | Flush heater core and | 30 | 00 |
| WI | Install heater valve and thermostat | 37 | 00 |
| JR | Install blower motor | 53 | 25 |
| RS | Install regulator | 16 | 50 |

| Comments: | | |
|---|---|---|
| TOTAL LABOR | | |
| TOTAL PARTS | | |
| 7% TAX ON PARTS | | |
| TOTAL | | |

**11.**

PARTS

| Quantity | Number | Description | Amount | |
|---|---|---|---|---|
| 1 | B9606 | alt. belt | 13 | 19 |
| | | | | |
| | | | | |
| | | | | |
| | | | | |
| | | | | |
| | | | | |
| | | | | |
| | | | | |
| | | | | |
| | | | | |
| | | | | |
| | | | | |
| | | | | |
| | | | | |
| | | | | |
| | | Total Parts | | |

CAR CLINIC
1257 Madison Street
Los Angeles, CA 90035-1257

CAR CLINIC

REPAIR ORDER

Name: Ms. Penny Owen Date: 5/15/- -
Address: 705 4 St. Phone: 277-6516

| Make & Model | Year | License | Mileage |
|---|---|---|---|
| Oldsmobile Ciera | 1988 | TGB-267 | 19,320 |

| Mechanic's Initials | LABOR Description | Amount | |
|---|---|---|---|
| RS | Check charging and system | 10 | 00 |
| RS | Replace alternator belt and charge battery | 16 | 00 |
| Comments: | TOTAL LABOR | | |
| | TOTAL PARTS | | |
| | 7% TAX ON PARTS | | |
| | TOTAL | | |

**12.**

PARTS

| Quantity | Number | Description | Amount | |
|---|---|---|---|---|
| 8 | S5729 | Spark plugs | 28 | 56 |
| 1 | F2155 | Fuel filter | 4 | 95 |
| 1 | F3284 | Air filter | 10 | 31 |
| 1 | F1121 | Crankcase filter | 3 | 50 |
| 1 | C6596 | Choke brake | 32 | 79 |
| | | | | |
| | | | | |
| | | | | |
| | | | | |
| | | | | |
| | | | | |
| | | | | |
| | | | | |
| | | | | |
| | | | | |
| | | | | |
| | | Total Parts | | |

CAR CLINIC
1257 Madison Street
Los Angeles, CA 90035-1257

CAR CLINIC

REPAIR ORDER

Name: Charles Shugart Date: 5/15/- -
Address: 2906 53 St. Phone: 275-9144

| Make & Model | Year | License | Mileage |
|---|---|---|---|
| Jeep Wagoneer | 1990 | T2A-994 | 9,885 |

| Mechanic's Initials | LABOR Description | Amount | |
|---|---|---|---|
| JR | Minor tune up | 41 | 25 |
| RS | Replace filters | 8 | 00 |
| Comments: | TOTAL LABOR | | |
| | TOTAL PARTS | | |
| | 7% TAX ON PARTS | | |
| | TOTAL | | |

# REPAIR ORDER TOTALS

Mr. Horton says to you, "Please get the total of all the repair orders. We call this the 'grand' total. I need this amount to review the May orders at the end of the month. I will also need the totals for labor, parts, and sales tax for May. Use the Repair Order Totals Form."

You answer, "I understand how important it is to have these amounts to see how well the business is doing."

## What You Are To Do

1. Calculate the Grand Total. Add the TOTAL amounts for each repair order to get the Grand Total. Record your answer in the Grand Total space at the top of the Repair Order Totals form.

CAR CLINIC
1257 Madison Street
Los Angeles, CA 90035-1257

REPAIR ORDER TOTALS
Week of May 11–15

| | |
|---|---|
| Grand Total | $ |
| | |
| Total Labor | $ |
| Total Parts | $ |
| Total Tax on Parts | $ |
| Grand Total | $ |

2. Calculate Total Labor. Add the Total Labor amounts for each repair order and record your answer on the Repair Order Totals form. Press [M+] to add this total to the memory.
3. Calculate Total Parts. Repeat step 1 for the Total Parts amount.
4. Calculate Total Tax. Repeat step 1 for the Total Tax on Parts amount.
5. Recall the Grand Total. Press [$M^R_C$] to find the Grand Total. (*Hint:* Check the Grand Total by comparing it with the

Grand Total at the top of the form. If the amounts are the same, your work is correct. If the amounts are not the same, repeat the steps until the two answers are the same.) Record your answer in the Grand Total space at the bottom of the Repair Order Totals form.

### Checking Your Work

Check your work with the answers at the back of the book. Record the number of correct answers on the Personal Progress Record. Correct the answers you got wrong by adding the amounts another time. Check your work again before you go on to the next step.

## MECHANICS' LABOR TOTALS

Mr. Horton asks, "Will you calculate the total amount of labor for each mechanic? I need these totals to figure their commissions. Use the Mechanics' Labor Totals form."

You answer, "Yes. I can see that this is important to you and to the mechanics so that they receive the correct pay."

### What You Are To Do

1. Clear your calculator's memory.
2. Calculate the total for William Ives. Add the labor amounts that go with WI (William Ives) in the Mechanic's Initials column on the repair orders. (WI will not appear on every repair order since William Ives did not repair every car.) Record your answer on the Mechanics' Labor Totals form. Press [M+] to add this total to the memory.

MECHANICS' LABOR TOTALS
Week of May 11–15

| Mechanic | Amount of Labor |
|---|---|
| William Ives (WI) | $ |
| Teri Rose (TR) | $ |
| Richard Shaw (RS) | $ |
| Total Labor | $ |

3. Calculate the total for Teri Rose. Repeat step 2 for the amounts marked TR.
4. Calculate the total for Richard Shaw. Repeat step 2 for the amounts marked RS.
5. Recall the total labor. Press [$M^R_C$] to find the Total Labor. (*Hint:* Compare this Total Labor with the Total Labor on the Repair Order Totals form on page 000.) Record your answer in the Total Labor space on the Mechanics' Labor Totals form.

### Checking Your Work

Check your work with the answers at the back of the book. Record the number of correct answers on the Personal Progress Record. Correct any answers you got wrong.

## REPAIR ORDER ANALYSIS

Mr. Horton studies a Repair Order Analysis form each week to see how he can improve his business. He asks you to complete the four parts of the form: Average Per Order; Percentage of Labor, Parts, and Tax; Percentage Difference Between Labor and Parts; and Percentage of Labor for Each Mechanic.

### What You Are to Do—Average Per Order

1. Copy the Grand Total for all of the repair orders (from the Repair Order Totals form) to the Repairs, Amount space on the Repair Order Analysis form.
2. Enter this same amount into your calculator.
3. Divide by the total number of repair orders (12). Round your answer to two decimal places.
4. Record your answer in the Repairs, Average space.
5. Copy the Total for all of the labor used to repair the cars (from the Repair Order Totals form) to the Labor, Amount space on the Repair Order Analysis form.
6. Enter this same amount into your calculator.
7. Divide by the total number of repair orders (12). Round your answer to two decimal places.
8. Record your answer in the Labor, Average space.
9. Repeat steps 5–8 for the Parts and Tax averages.

*Note:* Do not try to check your work by adding the averages for the labor, parts, and tax to see if the sum equals the average of the repairs. Remember that you rounded each of the averages to the nearest cent; therefore the sum may be off by a few cents.

| REPAIR ORDER ANALYSIS<br>Week of May 11–15 | | |
|---|---|---|
| **Average Per Order** | | |
| | Amount | Average |
| Repairs | $ | $ |
| Labor | $ | $ |
| Parts | $ | $ |
| Tax | $ | $ |
| **Percentage of Labor, Parts, and Tax** | | |
| | Amount | Percentage |
| Labor | $ | % |
| Parts | $ | % |
| Tax | $ | % |
| Total | $ | % |
| **Percentage Difference Between Labor and Parts** | | |
| | | _______ % |
| **Percentage of Labor for Each Mechanic** | | |
| | Amount | Percentage |
| William Ives (WI) | $ | % |
| Teri Rose (TR) | $ | % |
| Richard Shaw (RS) | $ | % |
| Total | $ | % |

## What You Are to Do—Percentage of Labor, Parts, and Tax

1. Copy the Total Labor amount from the Repair Order Totals form to the Labor, Amount space on the Repair Order Analysis form ($1,156.13).
2. Divide this Amount by the Grand Total from the Repair

Order Totals form to get the percentage. Round your answer to two decimal places.

3. Record your percentage in the Labor, Percentage space.
4. Repeat steps 1–3 for the Parts and Tax.
5. Total the Amount column and record your answer in the Total, Amount space.
6. Total the Percentage column and record your answer in the Total, Percentage space. The grand total ($2,405.97) is 100% of the total amount of labor, parts, and tax, so your total percentage should be 100%.

## What You Are To Do—Percentage Difference Between Labor and Parts

1. Subtract the Labor, Percentage from the Parts, Percentage.
2. Record your answer in the Percentage Difference Between Labor and Parts space.

## What You Are To Do—Percentage of Labor for Each Mechanic

1. Copy the Amount of Labor for William Ives (from the Mechanics' Labor Totals form) to the William Ives, Amount space on the Repair Order Analysis form.
2. Divide the Amount of Labor for William Ives by the Total Labor (from the Mechanics' Labor Totals form).
3. Change the decimal amount to a percentage. Round to two places.
4. Record your percentage in the William Ives, Percentage space.
5. Repeat steps 1–4 for the other two mechanics.
6. Total the Amount column and record your answer in the Total, Amount space ($1,156.13).
7. Total the Percentage column and record your answer in the Total, Percentage space. Your answer should be 100%.

## Checking Your Work

Check your work with the answers at the back of the book. Record the number of correct answers on the Personal Progress Record. Correct any answers you got wrong by repeating the steps above.

# WHAT YOU HAVE LEARNED

Congratulations on completing this work for Mr. Horton! You learned how important it is to be careful and accurate. Your estimating skills helped you know when you should refigure some totals. By using your calculator math skills you made sure customers were billed fairly and mechanics were given credit for the right amount of work. You also provided Mr. Horton with the information that he needed to see how the business was doing.

Mr. Horton tells you, "Thanks for the help. You did a good job."

You say, "Thank you for the chance. I enjoy working with the calculator."

# GLOSSARY

## A

**Addends** Two or more numbers to be added.
**Addition** Combining two or more numbers to get a total.
**Align** Place in a straight line.
**Average** Number found by totaling the addends and dividing by the number of addends in the problem.

## C

**Chain Division** Dividing a quotient by another divisor in a continuous process.
**Combined operations** Using two or more operations (addition, subtraction, multiplication, or division) to solve a problem.

## D

**Denominator** The number below the line in a fraction $(\frac{2}{[5]})$.
**Deposit** Money added to a checking account.
**Difference** The answer to a subtraction problem.
**Digit** One of the numbers 0, 1, 2, 3, 4, 5, 6, 7, 8, or 9.
**Dividend** The number to be divided in a division problem.
**Divisor** The number that the dividend is divided by.

## E

**Estimate** An approximate calculation.

## F

**Factors** The multiplicand and the multiplier in a multiplication problem.
**Fractions** Whole numbers separated into equal parts $(\frac{3}{4})$.

## H

**Horizontal** Across.

## I

**Improper Fraction** A fraction with the numerator greater than the denominator $(\frac{4}{3})$.
**Inventory** A complete list of the number and type of each item in stock.

## L

**Legibly** Clearly.

## M

**Memory Minus Key** [M−] Stores into memory a number which can be subtracted *from* another number in a problem.
**Memory Plus Key** [M+] Stores into memory a number which can be *added* to another number in a problem.
**Memory Recall and Clear Key** [$M^{R}_{C}$] If pressed once, brings a number from memory back into the answer window; if pressed twice, clears the calculator's memory.
**Minuend** The number being subtracted from.
**Mixed Number** A whole number (3) and a fraction $(\frac{1}{2})$. For example, $3\frac{1}{2}$.
**Multiplicand** The number in a multiplication problem which is added to itself a given number of times.
**Multiplication** Adding the multiplicand to itself the number of times given in the multiplier.
**Multiplier** The number in a multiplication problem which tells how many times to add the multiplicand to itself.

## N

**Numerator** The number above the line in a fraction $(\frac{[2]}{5})$.

## P

**Percent** Per hundred; shown by the symbol % (35%).

**Product** The answer to a multiplication problem.
**Proofread** (numbers) Compare each digit in a number with digits in another number to determine if they are identical.
**Proper Fraction** A fraction with the numerator less than the denominator ($\frac{4}{5}$).
**Prove the Answer** Work the problem a second time in a different way.

## Q

**Quotient** The answer to a division problem.

## R

**Round** To change a number to the nearest power of ten.

## S

**Subtraction** Deducting one number away from another number.
**Subtrahend** The number being subtracted.
**Sum** The answer to an addition problem; also called "total."

## T

**Total** The answer to an addition problem; also called "sum."

## V

**Vertical** Up and down.

## W

**Withdrawal** Money subtracted from a checking account.
**Word Problem** Math problem described in a story or scenario.

# INDEX

# ANSWERS

## UNIT 1

### CHECKPOINT 1-1, page 4

1. 83
2. 694
3. 372
4. 539
5. 970

### CHECKPOINT 1-2, page 5

1. seven two
2. three seven one
3. six nine eight
4. eight [pause] nine six four
5. two [pause] five one four

### CHECKPOINT 1-3, page 6

Answers will vary.

### CHECKPOINT 1-4, page 8

*Column C*
72HB609
6431 ÷ 839
11:08 a.m.
214-401-5970
8/14/88
368907

### ACTIVITY 1-1, page 10

1. 1,628
2. 1,637
3. 8,959
4. 99,772
5. 29,383

### ACTIVITY 1-2, page 10

1. $1,892.43
2. 12:38 p.m.
3. SQ-178-3062
4. 9/24/93
5. 15973628

## UNIT 2

### CHECKPOINT 2-1, page 12

1. 330
2. 5,100
3. 3,000
4. 7,600
5. 5,000

### CHECKPOINT 2-2, page 13

| | Rounded Addends | Est. Total | Calc. Total |
|---|---|---|---|
| 1. | 20 + 40 | 60 | 58 |
| 2. | 30 + 700 | 100 | 774 |
| 3. | 300 + 400 | 700 | 697 |
| 4. | 2,000 + 5,000 | 7,000 | 7,222 |
| 5. | 2,000 + 50,000 + 7,000 | 59,000 | 55,781 |

### CHECKPOINT 2-3, page 15

1. <u>all together</u> 33
2. No +
3. <u>total</u> $7,230
4. No +
5. <u>how much</u> $30

### CHECKPOINT 2-4, page 17

1. 43.52
2. $101.61
3. $740.59
4. 151.923
5. 45.7249

### ACTIVITY 2-1, page 19

1. 30
2. 80
3. 800
4. 4,000
5. 9,000

### ACTIVITY 2-2, page 19

| | Rounded Addends | Est. Total | Calc. Total |
|---|---|---|---|
| 1. | 40 + 50 | 90 | 89 |
| 2. | 50 + 200 | 250 | 244 |
| 3. | 600 + 200 | 800 | 855 |
| 4. | 800 + 900 | 1,700 | 1,723 |
| 5. | 100 + 600 | 700 | 701 |

### ACTIVITY 2-3, page 19

1. <u>in all</u> 1,030
2. 400
3. <u>all together</u> 100
4. <u>total</u> 139
5. <u>total</u> $3.40

### ACTIVITY 2-4, page 20

1. 79.45
2. $59.68
3. $48.31
4. 1,323.2719
5. 727.186

## UNIT 3

### CHECKPOINT 3-1, page 23

1. 21 Proof: 21 + 14 = 35
2. 45 Proof: 45 + 34 = 79

**3.** 254 Proof: 254 + 89 = 343

**4.** 331 Proof: 331 + 231 = 562

**5.** 6,574 Proof: 6,574 + 856 = 7,430

## CHECKPOINT 3-2, page 25

| | Rounded Problem | Estimated Difference | Calculator Difference |
|---|---|---|---|
| **1.** | 80 − 40 | 40 | 39 |
| **2.** | 900 − 80 | 820 | 786 |
| **3.** | 200 − 100 | 100 | 92 |
| **4.** | 400 − 200 | 200 | 205 |
| **5.** | 300 − 100 | 200 | 138 |

## CHECKPOINT 3-3, page 26

**1.** difference 8
**2.** less 194
**3.** reduces $0.56
**4.** remains $9
**5.** left 7

## CHECKPOINT 3-4, page 27

**1.** $56.30
**2.** 25.1054
**3.** 10.161
**4.** 1633.371
**5.** 144.18

## CHECKPOINT 3-5, page 28

**1.** 5.37
**2.** $24.95
**3.** 9.109
**4.** $10.56
**5.** 52.777

## JUST FOR FUN, page 28

**1.** 57,716 GILLS
**2.** 77,345,663 EGGSHELL
**3.** 38,076 GLOBE
**4.** 5,514 HISS
**5.** 818 BIB

## ACTIVITY 3-1, page 30

**1.** 35 Proof: 35 + 54 = 89

**2.** 359 Proof: 359 + 96 = 455

**3.** 75 Proof: 75 + 73 = 148

**4.** 639 Proof: 639 + 36 = 675

**5.** 3331 Proof: 3331 + 245 = 3576

## ACTIVITY 3-2, page 30

| | Rounded Problem | Estimated Difference | Calculator Difference |
|---|---|---|---|
| **1.** | 50 − 20 | 30 | 31 |
| **2.** | 200 − 50 | 150 | 188 |
| **3.** | 500 − 100 | 400 | 372 |
| **4.** | 200 − 60 | 140 | 122 |
| **5.** | 400 − 300 | 100 | 118 |

## ACTIVITY 3-3, page 31

**1.** difference $56
**2.** change $0.27
**3.** less $75
**4.** left over $2.64
**5.** remains $169.31

## ACTIVITY 3-4, page 31

**1.** 19.451
**2.** $338.55
**3.** 59385.673
**4.** $60.97
**5.** 340.375

## REVIEW 3-1, page 32

| | |
|---|---|
| Pants | $18.04 |
| Socks | $2.81 |
| Shirt | $4.74 |
| Sweater | $5.50 |
| *Totals* | |
| Regular Price | $134.35 |
| Sale Price | $95.86 |
| Difference | $38.49 |

## REVIEW 3-2, page 33

| | |
|---|---|
| Bread | $1.19 |
| Dog Food | $7.24 |
| Milk | $2.40 |
| Ground Beef | $1.88 |
| *Totals* | |
| Regular Price | $18.09 |
| Less Coupon | $3.88 |
| Cost with Coupon | $14.21 |

## UNIT 4

### CHECKPOINT 4-1, page 35

1. 294
2. 59,644
3. 1,371,905
4. 2,945
5. 227,752

### CHECKPOINT 4-2, page 36

1. 135
2. No ×
3. 405
4. No ×
5. 124

### CHECKPOINT 4-3, page 38

| | Unrounded Product | Rounded Product |
|---|---|---|
| 1. | 59.5886 | 59.59 |
| 2. | 5.912286 | 5.912 |
| 3. | 51.596877 | 51.597 |
| 4. | 25.27084 | 25.2708 |
| 5. | 27.052577 | 27.0526 |

### CHECKPOINT 4-4, page 39

1. $13.31
2. $63.80
3. $155.13
4. $359.80
5. $5,400.00

### CHECKPOINT 4-5, page 40

| | Rounded Factors | Estimated Product | Calculator Product |
|---|---|---|---|
| 1. | 90 × 30 | 2,700 | 2,752 |
| 2. | 400 × 60 | 24,000 | 24,095 |
| 3. | 500 × 300 | 150,000 | 140,756 |
| 4. | 4,000 × 50 | 200,000 | 195,216 |
| 5. | 9,000 × 400 | 3,600,000 | 3,665,458 |

### JUST FOR FUN, page 40

1. 7,738 Bell
2. 338 Bee
3. 7,714 Hill
4. 604 Hog

### ACTIVITY 4-1, page 42

1. 5,330
2. 46,797
3. 9,746,480
4. 49,776
5. 4,533,056

### ACTIVITY 4-2, page 42

1. 350
2. 480
3. 1,860
4. 3,750
5. 3,150

### ACTIVITY 4-3, page 43

| | Unrounded Product | Rounded Product |
|---|---|---|
| 1. | 0.10608 | 0.11 |
| 2. | 0.062152 | 0.062 |
| 3. | 320.0262 | 320.026 |
| 4. | 124.51146 | 124.5115 |
| 5. | 295.29528 | 295.2953 |

### ACTIVITY 4-4, page 43

1. $21.76
2. $68.85
3. $195.56
4. $333.30
5. $875.00

### ACTIVITY 4-5, page 44

| | Rounded Factors | Estimated Product | Calculator Product |
|---|---|---|---|
| 1. | 70 × 100 | 7,000 | 7,104 |
| 2. | 500 × 90 | 45,000 | 44,370 |
| 3. | 400 × 200 | 80,000 | 84,000 |
| 4. | 2,000 × 100 | 200,000 | 220,388 |
| 5. | 7,000 × 4,000 | 28,000,000 | 31,893,585 |

### REVIEW 4-1, page 45

Tomatoes, 180
Cucumbers, 44
Carrots, 108
Cantelope, 38
Corn, 126
Onions, 204

## UNIT 5

### CHECKPOINT 5-1, page 47

1. 26
2. 182
3. 153
4. 248
5. 26

### CHECKPOINT 5-2, page 48

1. 6
2. No ÷
3. 20
4. No ÷
5. 400

### CHECKPOINT 5-3, page 50

| | Unrounded Quotient | Rounded Quotient |
|---|---|---|
| 1. | 9.875 | 9.88 |
| 2. | 12.0375 | 12.038 |
| 3. | 48.428571 | 48.429 |
| 4. | 18.357142 | 18.3571 |
| 5. | 4.4313725 | 4.4314 |

## CHECKPOINT 5-4, page 51

| | Rounded Numbers | Est. Quo. | Calc. Quo. |
|---|---|---|---|
| **1.** | 200 ÷ 20 | 10 | 8 |
| **2.** | 2,000 ÷ 50 | 40 | 34 |
| **3.** | 3,000 ÷ 30 | 100 | 84 |
| **4.** | 5,000 ÷ 300 | 17 | 14 |
| **5.** | 30,000 ÷ 1,000 | 30 | 23 |

## CHECKPOINT 5-5, page 51

**1.** $0.04
**2.** $2.10
**3.** $131.25
**4.** $0.06
**5.** $1.98

## CHECKPOINT 5-6, page 52

**1.** 1.41
**2.** 1.87
**3.** 7
**4.** 231.86
**5.** 26

## JUST FOR FUN, page 53

**1.** 278,804 HOBBLE
**2.** 7,108 BOIL
**3.** 378,806 GOBBLE
**4.** 3,045 SHOE
**5.** 3,704 HOLE

## ACTIVITY 5-1, page 54

**1.** 21.25
**2.** 38.375
**3.** 35.5
**4.** 7
**5.** 34

## ACTIVITY 5-2, page 54

**1.** 80
**2.** 50
**3.** 21
**4.** 17
**5.** 9

## ACTIVITY 5-3, page 55

| | Rounded Numbers | Estimated Quotient | Calculator Quotient |
|---|---|---|---|
| **1.** | 200 ÷ 10 | 20 | 47 |
| **2.** | 600 ÷ 50 | 12 | 13 |
| **3.** | 6,000 ÷ 10 | 600 | 634 |
| **4.** | 3,000 ÷ 20 | 150 | 223 |
| **5.** | 50,000 ÷ 700 | 71.43 or 71 | 69 |

## ACTIVITY 5-4, page 55

**1.** $9.85
**2.** $4.75
**3.** $2.50
**4.** 730
**5.** $34.60

## ACTIVITY 5-5, page 56

| | Unrounded Quotient | Rounded Quotient |
|---|---|---|
| **1.** | 9.075 | 9.08 |
| **2.** | 95.333333 | 95.333 |
| **3.** | 1.6875 | 1.688 |
| **4.** | 6.6301369 | 6.6301 |
| **5.** | 2.2123456 | 2.2123 |

## ACTIVITY 5-6, page 57

**1.** 9
**2.** 0.5
**3.** 59
**4.** 5
**5.** 256

# UNIT 6

## CHECKPOINT 6-1, page 60

**1.** 77.5
**2.** 26.25
**3.** 82.75
**4.** 178
**5.** 43.6

## CHECKPOINT 6-2, page 62

**1.** $1,680
**2.** $15.07
**3.** $5.25
**4.** 7
**5.** $136

## CHECKPOINT 6-3, page 64

**1.** 10
**2.** $155.86
**3.** 20
**4.** 2.16
**5.** 0.4375

## JUST FOR FUN, page 65

**1.** 6 = G
**2.** 461,375 = SLEIGH
**3.** 376,616 = GIGGLE

## CHECKPOINT 6-4, page 66

| | Rounded Problem | Est. Ans. | Calc. Answer |
|---|---|---|---|
| **1.** | 200 × 3 + 20 | 620 | 662 |
| **2.** | 50 × .2 + 50 | 60 | 57.576 |
| **3.** | 20 + 10 ÷ 2 | 15 | 15.023474 |
| **4.** | 50 + 80 + 80 ÷ 3 | 70 | 69 |
| **5.** | 2,000 − 90 − 200 + 20 | 1,730 | 1,752 |

## ACTIVITY 6-1, page 68

**1.** 2.4
**2.** 554
**3.** 271.25
**4.** 274
**5.** 60

## ACTIVITY 6-2, page 68

**1.** $323
**2.** 60
**3.** $477.42
**4.** $200
**5.** 16

## ACTIVITY 6-3, page 69

**1.** 157
**2.** $34.31
**3.** 1895.1428
**4.** 3.15
**5.** 86.8

## ACTIVITY 6-4, page 70

| | Rounded Problem | Est. Ans. | Calc. Ans. |
|---|---|---|---|
| **1.** | 6 + 3 + 3 ÷ 3 | 4 | 3.9 |
| **2.** | 500 + 60 − 40 − 90 | 430 | 484 |
| **3.** | 8 + 2 ÷ 3 | 3.33 | 4 |
| **4.** | 17 × .1 + 17 | 18.7 | 18.1365 |
| **5.** | 90 × 4 + 100 ÷ 5 | 92 | 89.6 |

## REVIEW 6-1, page 71

Checkbook Register

| *Date* | *Description* | *Deposit* | *With-drawal* | *Balance* |
|---|---|---|---|---|
| 12/1 | Beginning Balance | | | $342.61 |
| 12/3 | Pam Gared (Rent) | | $250 | $97.61 |
| 12/10 | Kinu Oswan | $35 | | $127.61 |
| 12/20 | Food Day Groceries | | $75.56 | $52.05 |
| 12/29 | Gears (paycheck) | $335.50 | | $387.55 |
| 12/30 | Everelec (elec. bill) | | $27.86 | $359.69 |

What is the ending balance? $359.69

## REVIEW 6-2, page 71

CARPENTER MEDICAL CENTER
Attendant's Time Record

*Patient*

Name Julia McCentric
Room #534 Phone (605) 555-7239

*Attendant*

Name Kelly Carter
Address Blackwood Village Apt. #507
Phone (605) 555-8215
Social Security # 346-58-7372

| Date of Service | Hours Worked | Hours |
|---|---|---|
| 7/3 | 7:30–12:30 | 5 |
| 7/6 | 7:30–10:30 | 3 |
| 7/9 | 8:00–12:00 | 4 |
| 7/12 | 7:30–11:30 | 4 |
| 7/15 | 7:00–12:00 | 5 |

Total Hours 21
Hourly Rate $6.00
Gross Salary $126
Signature ____________
Approved by Administrator ____________

# UNIT 7

## CHECKPOINT 7-1, page 75

**1.** 2,099
**2.** 1,989
**3.** 1,611
**4.** 2,170
**5.** Grand Total = 8,983

## CHECKPOINT 7-2, page 77

**1.** $15
**2.** $463.38
**3.** $14.09
**4.** Marton $7.83
**5.** $.03

## ACTIVITY 7-1, page 80

**1.** 1,940
**2.** 2,505
**3.** 1,516
**4.** 1,281
**5.** Grand Total = 9,079

## ACTIVITY 7-2, page 80

**1.** $27.82
**2.** $189.62
**3.** Generic $0.05
**4.** $213.77
**5.** 7,320

## REVIEW 7-1, page 83

| | Amount Due |
|---|---|
| AQ477 | $ 597.90 |
| AP3646 | $ 755.88 |
| QDAP-13 | $ 153.65 |
| CPQ-62 | $ 191.76 |
| AP-3615 | $ 129.75 |
| CPQ-65 | $ 71.40 |
| SUBTOTAL | $1900.34 |
| TAX | $ 142.53 |
| TOTAL | $2042.87 |

## REVIEW 7-2, page 84

| Inventory Price |
|---|
| 124.75 |
| 49.90 |
| 104.79 |
| 279.44 |
| 59.33 |
| 55.84 |
| 66.31 |
| 181.48 |
| 27.59 |
| 27.59 |
| 27.59 |
| 82.77 |
| 27.80 |
| 34.75 |
| 62.55 |
| 606.24 |

# UNIT 8

## CHECKPOINT 8-1, page 90

**1.** proper
**2.** mixed
**3.** improper
**4.** proper
**5.** mixed

## CHECKPOINT 8-2, page 91

**1.** .875
**2.** .464
**3.** .875
**4.** .493
**5.** .787

## CHECKPOINT 8-3, page 93

**1.** 8.25 or $8\frac{1}{4}$ cups
**2.** $252
**3.** $285.86
**4.** $7.50
**5.** 27.25 or $27\frac{1}{4}$

## ACTIVITY 8-1, page 95

**1.** proper
**2.** improper
**3.** mixed
**4.** improper
**5.** mixed

## ACTIVITY 8-2, page 95

**1.** .726
**2.** 1.061
**3.** .630
**4.** .635
**5.** 2.605

## ACTIVITY 8-3, page 95

**1.** $1.33
**2.** $.52
**3.** $.46
**4.** $5.00
**5.** 7:00-8:00 55.8
8:00-9:00 57.13

## REVIEW 8-1, page 97

Invoice

| | |
|---|---|
| 2 Birthday Cakes | $33.46 |
| $3\frac{1}{4}$ dozen Peanut butter cookies | $ 3.19 |
| TOTAL | $36.65 |

## REVIEW 8-2, page 98

| Taken |
|---|
| |

| Amount |
|---|
| .31 |
| |
| 1.22 |
| |
| 1.61 |
| |
| |
| 3.14 |
| |

# UNIT 9

## CHECKPOINT 9-1, page 102

**1.** $48.09
**2.** $46.20
**3.** $75.09
**4.** $183.84
**5.** $6.28
**6.** $33.46
**7.** $165.38
**8.** $304.44
**9.** $60.15
**10.** $69.21

## JUST FOR FUN, page 104

The answer is always the same as the number entered in step 1.

## CHECKPOINT 9-2, page 104

**1.** 14.5%
**2.** 30%
**3.** 17.4%
**4.** 24.7%
**5.** 25.8%

## CHECKPOINT 9-3, page 105

**1.** 23.62%
**2.** 2.74%
**3.** 12.28%
**4.** 20.08%
**5.** 7%

## ACTIVITY 9-1, page 108

**1.** $11,351
**2.** $15,135
**3.** $38,664
**4.** $17,618
**5.** $12,415

## ACTIVITY 9-2, page 109

**1.** $18.86
**2.** 53.70%
**3.** $11.44
**4.** 19.75%
**5.** $6.50
**6.** 22.82%
**7.** $5.74
**8.** 40.34%
**9.** $5.98
**10.** 38.16%

## ACTIVITY 9-3, page 110

**1.** 7.51%
**2.** 15.34%
**3.** 14.29%
**4.** $312.50
**5.** 54.84%

# PART 4

## REPAIR ORDERS 1–12, pages 116–121

| Repair Order # | Total Labor | Total Parts | Tax on Parts | Total |
|---|---|---|---|---|
| **1** | $391.25 | $163.39 | $11.44 | $566.08 |
| **2** | 37.50 | 107.43 | 7.52 | 152.45 |
| **3** | 162.75 | 102.64 | 7.18 | 272.57 |
| **4** | 125.75 | 50.53 | 3.54 | 179.82 |
| **5** | 101.75 | 163.38 | 11.44 | 276.57 |
| **6** | 10.00 | .89 | .06 | 10.95 |
| **7** | 13.50 | 65.72 | 4.60 | 83.82 |
| **8** | 54.75 | 146.43 | 10.25 | 211.43 |
| **9** | 46.88 | 84.67 | 5.93 | 137.48 |
| **10** | 136.75 | 189.69 | 13.28 | 339.72 |
| **11** | 26.00 | 13.19 | .92 | 40.11 |
| **12** | 49.25 | 80.11 | 5.61 | 134.97 |

## REPAIR ORDER TOTALS, page 122

| | |
|---|---|
| Grand Total | $2,405.97 |
| Total Labor | 1,156.13 |
| Total Parts | 1,168.07 |
| Total Tax on Parts | 81.77 |
| Grand Total | 2,405.97 |

## MECHANICS' LABOR TOTALS, page 123

| Mechanic | Amount of Labor |
|---|---|
| William Ives (WI) | $ 122.25 |
| Teri Rose (TR) | 529.13 |
| Richard Shaw (RS) | 504.75 |
| Total Labor | 1,156.13 |

## REPAIR ORDER ANALYSIS, page 125

Average Per Order

| | Amount | Average |
|---|---|---|
| Repairs | $2,405.97 | $200.50 |
| Labor | 1,156.13 | 96.34 |
| Parts | 1,168.07 | 97.34 |
| Tax | 81.77 | 6.81 |

Percentage of Labor, Parts, and Tax

| | Amount | Percentage |
|---|---|---|
| Labor | $1,156.13 | 48.05% |
| Parts | 1,168.07 | 48.55% |
| Tax | 81.77 | 3.40% |
| Total | 2,405.97 | 100.00% |

Percentage Difference Between Labor and Parts

.50%

Percentage of Labor for Each Mechanic

| | Amount | Percentage |
|---|---|---|
| William Ives (WI) | $ 122.25 | 10.57% |
| Teri Rose (TR) | 529.13 | 45.77% |
| Richard Shaw (RS) | 504.75 | 43.66% |
| Total | 1,156.13 | 100.00% |

# PERSONAL PROGRESS RECORD

Name: ______________________________

To record your score:
1. Count how many problems you have correct.
2. Record the number correct in the Score column.

### UNIT 1: Addition—The Basics

| Exercise | Score |
|---|---|
| Checkpoint 1-1 | |
| Checkpoint 1-2 | |
| Checkpoint 1-3 | |
| Checkpoint 1-4 | |
| Activity 1-1 | |
| Activity 1-2 | |
| Total | |

**HOW ARE YOU DOING?**

| | |
|---|---|
| **Better than 26** | **Excellent** |
| **21–25** | **Good** |
| **19–20** | **Fair** |
| **Less than 19** | **See Instructor** |

### UNIT 2: Addition—Using It Every Day

| Exercise | Score |
|---|---|
| Checkpoint 2-1 | |
| Checkpoint 2-2 | |
| Checkpoint 2-3 | |
| Checkpoint 2-4 | |
| Activity 2-1 | |
| Activity 2-2 | |
| Activity 2-3 | |
| Activity 2-4 | |
| Total | |

**HOW ARE YOU DOING?**

| | |
|---|---|
| **Better than 48** | **Excellent** |
| **42–47** | **Good** |
| **36–41** | **Fair** |
| **Less than 36** | **See Instructor** |

### UNIT 3: Subtraction—Using It Every Day

| Exercise | Score |
|---|---|
| Checkpoint 3-1 | |
| Checkpoint 3-2 | |
| Checkpoint 3-3 | |
| Checkpoint 3-4 | |
| Checkpoint 3-5 | |
| Activity 3-1 | |
| Activity 3-2 | |
| Activity 3-3 | |
| Activity 3-4 | |
| Review 3-1 | |
| Review 3-2 | |
| Total | |

**HOW ARE YOU DOING?**

| | |
|---|---|
| **Better than 56** | **Excellent** |
| **46–55** | **Good** |
| **40–45** | **Fair** |
| **Less than 40** | **See Instructor** |

## UNIT 4: Multiplication—Using It Every Day

| Exercise | Score |
|---|---|
| Checkpoint 4-1 | |
| Checkpoint 4-2 | |
| Checkpoint 4-3 | |
| Checkpoint 4-4 | |
| Checkpoint 4-5 | |
| Activity 4-1 | |
| Activity 4-2 | |
| Activity 4-3 | |
| Activity 4-4 | |
| Activity 4-5 | |
| Review 4-1 | |
| Total | |

**HOW ARE YOU DOING?**

| | |
|---|---|
| Better than 56 | Excellent |
| 49–55 | Good |
| 41–48 | Fair |
| Less than 41 | See Instructor |

## UNIT 5: Division—Using It Every Day

| Exercise | Score |
|---|---|
| Checkpoint 5-1 | |
| Checkpoint 5-2 | |
| Checkpoint 5-3 | |
| Checkpoint 5-4 | |
| Checkpoint 5-5 | |
| Checkpoint 5-6 | |
| Activity 5-1 | |
| Activity 5-2 | |
| Activity 5-3 | |
| Activity 5-4 | |
| Activity 5-5 | |
| Activity 5-6 | |
| Total | |

**HOW ARE YOU DOING?**

| | |
|---|---|
| Better than 72 | Excellent |
| 62–71 | Good |
| 54–61 | Fair |
| Less than 54 | See Instructor |

## UNIT 6: Combined Operations—Using Them Every Day

| Exercise | Score |
|---|---|
| Checkpoint 6-1 | |
| Checkpoint 6-2 | |
| Checkpoint 6-3 | |
| Checkpoint 6-4 | |
| Activity 6-1 | |
| Activity 6-2 | |
| Activity 6-3 | |
| Activity 6-4 | |
| Review 6-1 | |
| Review 6-2 | |
| Total | |

**HOW ARE YOU DOING?**

| | |
|---|---|
| Better than 58 | Excellent |
| 49–57 | Good |
| 42–49 | Fair |
| Less than 42 | See Instructor |

## UNIT 7: Memory Operations—Using Them Every Day

| Exercise | Score |
|---|---|
| Checkpoint 7-1 | ________ |
| Checkpoint 7-2 | ________ |
| Activity 7-1 | ________ |
| Activity 7-2 | ________ |
| Review 7-1 | ________ |
| Review 7-2 | ________ |
| Total | ________ |

**HOW ARE YOU DOING?**

| | |
|---|---|
| Better than 37 | Excellent |
| 29–36 | Good |
| 25–28 | Fair |
| Less than 25 | See Instructor |

## UNIT 8: Fractions and Decimals—Using Them Every Day

| Exercise | Score |
|---|---|
| Checkpoint 8-1 | ________ |
| Checkpoint 8-2 | ________ |
| Checkpoint 8-3 | ________ |
| Activity 8-1 | ________ |
| Activity 8-2 | ________ |
| Activity 8-3 | ________ |
| Review 8-1 | ________ |
| Review 8-2 | ________ |
| Total | ________ |

**HOW ARE YOU DOING?**

| | |
|---|---|
| Better than 32 | Excellent |
| 27–31 | Good |
| 23–26 | Fair |
| Less than 23 | See Instructor |

## UNIT 9: Percents—Using Them Every Day

| Exercise | Score |
|---|---|
| Checkpoint 9-1 | ________ |
| Checkpoint 9-2 | ________ |
| Checkpoint 9-3 | ________ |
| Activity 9-1 | ________ |
| Activity 9-2 | ________ |
| Activity 9-3 | ________ |
| Total | ________ |

**HOW ARE YOU DOING?**

| | |
|---|---|
| Better than 34 | Excellent |
| 28–33 | Good |
| 24–27 | Fair |
| Less than 24 | See Instructor |

## SIMULATION

| Repair Orders | Score |
|---|---|
| 1 | ________ |
| 2–12 | ________ |
| Total | ________ |

**HOW ARE YOU DOING?**

| | |
|---|---|
| Better than 43 | Excellent |
| 38–42 | Good |
| 33–37 | Fair |
| Less than 33 | See Instructor |

*Note:* You must correct *every* answer you got wrong before continuing with the simulation. These answers are used for the rest of the simulation.

| | Score |
|---|---|
| Repair Order Totals | ______ |
| Mechanics' Labor Totals | ______ |
| Total | ______ |

**HOW ARE YOU DOING?**

| | |
|---|---|
| **8** | **Excellent** |
| **7** | **Good** |
| **6** | **Fair** |
| **Less than 6** | **See Instructor** |

*Note:* Don't forget to correct any answers you got wrong.

| | Score |
|---|---|
| Repair Order Analysis | ______ |

**HOW ARE YOU DOING?**

| | |
|---|---|
| **Better than 19** | **Excellent** |
| **17** | **Good** |
| **15** | **Fair** |
| **Less than 15** | **See Instructor** |